Theory of Digital Automata

International Series on
INTELLIGENT SYSTEMS, CONTROL AND AUTOMATION:
SCIENCE AND ENGINEERING

VOLUME 63

For further volumes:
http://www.springer.com/series/6259

Bohdan Borowik • Mykola Karpinskyy
Valery Lahno • Oleksandr Petrov

Theory of Digital Automata

Springer

Bohdan Borowik
Department of Electrical Engineering
University of Bielsko-Biala
Bielsko-Biala, Poland

Mykola Karpinskyy
Computer Science Division
University of Bielsko-Biala
Bielsko-Biala, Poland

Valery Lahno
Department of Computer Systems and Networks
East-Ukrainian National University
Luhansk, Ukraine

Oleksandr Petrov
Department of Applied Computer Science
AGH University of Science and Technology
Krakow, Poland

Reviewer
Barbara Borowik
Cracow University of Technology/Faculty of Physics
Mathematics and Computer Science
Cracow, Poland

ISBN 978-94-007-5227-6 ISBN 978-94-007-5228-3 (eBook)
DOI 10.1007/978-94-007-5228-3
Springer Dordrecht Heidelberg New York London

Library of Congress Control Number: 2012951321

Printed on acid-free paper

Springer is part of Springer Science+Business Media (www.springer.com)

Preface

Communication systems for most companies throughout the world have already gone digital or will certainly do so in the near future. For example, cell phones and other types of wireless communication such as television, radio, process controls, automotive electronics, consumer electronics, global navigation and military systems, to name only a few applications, depend heavily on digital electronics.

This book is designed to serve as a first course in digital automata and digital systems, providing students at the sophomore level a transition from the world of physics to the world of digital electronics and computation.

The book attempts to satisfy two goals: Combine the study of circuits and digital electronics into a single, unified treatment, and establish a strong connection with the contemporary worlds of both these types of digital systems.

These goals arise from the observation that the conventional approach to introducing digital electronics through a course in traditional circuit analysis is fast becoming obsolete. Our world has gone digital. Even those students who remain in core electrical engineering are heavily influenced by the digital domain.

Because of this elevated focus on the digital domain, basic electrical engineering education must change in two ways: First, the traditional approach to teaching circuits and electronics without regard to the digital domain must be replaced by one that stresses the foundations that are common to all circuits in both the digital and analog domains. Because most of the fundamental concepts in circuits and electronics are equally applicable to both the digital and the analog domains, this means that, primarily, we must change the way in which we motivate the study of circuits and electronics to emphasize their broader impact on digital systems.

Second, given the additional demands of computer engineering, many departments can ill-afford the luxury of separate courses on circuits and on electronics. Rather, they might be combined into one course.

The book attempts to form a bridge between the world of logic and the world of large digital systems.

Contents

Introduction

Early artificial intelligence theory was concerned with models (automata) used to simulate objects and processes. Automata theory helps with the design of digital circuits such as parts of computers, telephone systems or control systems.

The major advantages for Digital and Microprocessor systems are [1–5]:

- Stability and accuracy of control;
- Flexibility;
- Lower cost per function;
- Greater reliability and equipment life;
- Human factors favouring a Digital Interface.

The most general and versatile circuit that can be placed on a single Chip is the Digital Microprocessor. The Microprocessor is versatile because it can be programmed to perform an almost unlimited number of computing tasks.

Application and Advantages of Digital Systems

- The device used in a Digital Circuit generally operates in one of the two states, known as ON & OFF, resulting in a very simple operation.
- There are only a few basic operations in a Digital Circuit and they are very easy to understand.
- Digital Technique requires Boolean algebra which is very simple and easily to learn.
- Digital Circuit study requires the basic concept of Electrical Network Analysis, which is also easily learned.
- A large number of Integrated Circuits (IC) are available for performing various operations. They are highly reliable and accurate, with a very high speed of operations.

- Digital Circuits have a wide range of memory capability which makes them highly suitable for Computers, Calculators, and Electronic Watches etc.
- The display of data and other information is very convenient, accurate and elegant using digital techniques.

Many students have, in a wide range of studies, opportunities to learn programming of digital computers, hence they have a strong motivation to study the way digital hardware works.

Chapter 1
Digital and Analog Quantities

Abstract This chapter discusses different digital representations commonly used to represent data. In electronic applications digital representations have certain advantages over analog representations. The chapter introduces also to analog and digital quantities and to Post–Turing machines.

1.1 Analog and Digital Quantities

Electronic circuits can be divided into two broad categories, digital and analog. Digital electronics involves quantities with discrete values, and analog electronics involves quantities with continuous values.

A digital quantity is one having a discrete set of values. Most things that can be measured quantitatively occur in nature in analog form. For example, $S(t) = f(t)$, Fig. 1.1.

Analog and digital quantities – An analog signal is sampled or tested repeatedly over a period of time to determine the characteristic that contains the analog quantity. The sampled analog value is converted to the nearest binary value or quantity. The binary value is then encoded into a character stream acceptable to the digital equipment that is designed to use the data. Standardized binary words called BAMs (binary angular measurement) are used to transmit angular, range, and height values between digital equipment in shipboard combat direction systems. Other coding systems such as Gray code or binary-coded decimal (BCD) are also used to transmit converted values.

Advantage Digital representation has certain advantages over analog representation in electronics applications. For one thing, digital data can be processed and transmitted more efficiently and reliably than analog data. Also, digital data has a great advantage when storage is necessary. For example, music when converted to digital form can be stored more compactly and reproduced with greater accuracy and clarity than is possible when it is in analog form. Noise (unwanted voltage fluctuations) does not affect digital data nearly as much as it does analog signals, Fig. 1.2.

B. Borowik et al., *Theory of Digital Automata*, Intelligent Systems, Control and Automation: Science and Engineering 63, DOI 10.1007/978-94-007-5228-3_1,

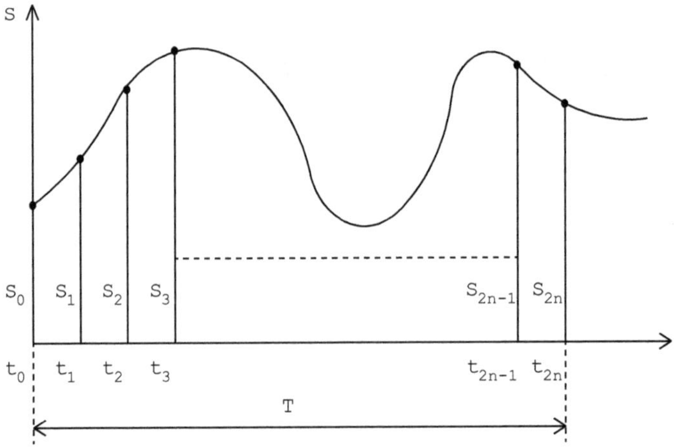

Fig. 1.1 Quantization of the analog quantity

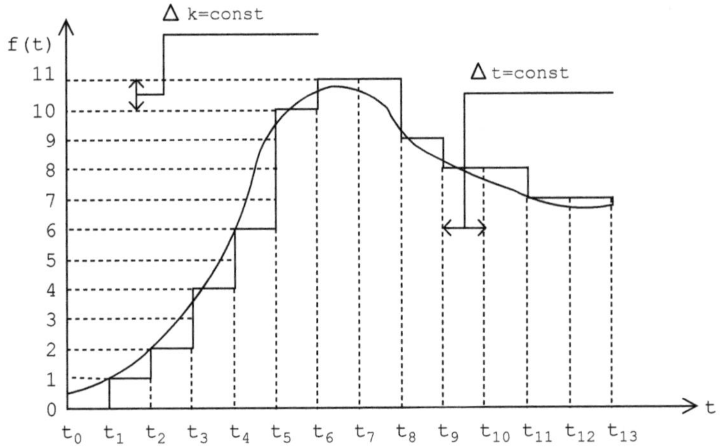

Fig. 1.2 Digital representation

Sampled-value representation (quantization) of the analog quantity in Fig. 1.1. Each value represented by a dot can be digitized by representing it as a digital code that consists of a series of 1s and 0s:

$$S(t) = \sum_{k=-\infty}^{\infty} S\left(t - k \cdot \Delta t\right) \cdot \frac{sin\left[2\pi \cdot F_m \cdot \left(t - k \cdot \Delta t\right)\right]}{2\pi \cdot F_m \cdot \left(t - k \cdot \Delta t\right)},$$

where $k = \left(f_{max} - f_{min}\right) / \left(q - 1\right)$;
$\Delta t = 1/(2F_m)$ – step quantization;
q – number step quantization;
$t_k = k/2 \cdot F_m = k \cdot \Delta t.$

Table 1.1 Analog versus digital quantities

	Analog signals	Digital signals
Technology	Analog technology records waveforms as they are.	Converts analog waveforms into set of numbers and records them. The numbers are converted into voltage stream for representation.
Representation analog and digital signals	Uses continuous range of values to represent information.	Uses discrete or discontinuous values to represent information.
Uses analog and digital signals	Can be used in various computing platforms and under operating systems.	Computing and electronics technology.
Computer	Analog computer uses changeable continuous physical phenomena such as electrical, mechanical, hydraulic quantities so as to solve a problem.	Digital computers represent changing quantities incrementally as and when their values change.

Analog and digital signals are used to transmit information, usually through electric signals. In both these technologies, the information, such as any audio or video, is transformed into electric signals. The difference between analog and digital technologies is that in analog technology, information is translated into electric pulses of varying amplitude. In digital technology, translation of information is into binary format (0 or 1) where each bit is representative of two distinct amplitudes, see Table 1.1 [4–6].

All digital information possesses common properties that distinguish it from analog communications methods [1, 4]:

Synchronization: Since digital information is conveyed by the sequence in which symbols are ordered, all digital schemes have some method for determining the beginning of a sequence.

Language: All digital communications require a language, which in this context consists of all the information that the sender and receiver of the digital communication must both possess, in advance, in order for the communication to be successful.

Errors: Disturbances (noise) in analog communications invariably introduce some, generally small, deviation or error between the intended and actual communication. Disturbances in a digital communication do not result in errors unless the disturbance is so large as to result in a symbol being misinterpreted as another symbol or disturb the sequence of symbols. It is therefore generally possible to have an entirely error-free digital communication. Further, techniques such as check codes may be used to detect errors and guarantee error-free communications through redundancy or retransmission. Errors in digital communications can take the form of substitution errors in which a symbol is replaced by another symbol, or insertion/deletion errors in which an extra incorrect symbol is inserted into or deleted from a digital message. Uncorrected errors in digital communications have unpredictable and generally large impact on the information content of the communication.

Granularity: When a continuously variable analog value is represented in digital form there is always a decision as to the number of symbols to be assigned to that value. The number of symbols determines the precision or resolution of the resulting datum. The difference between the actual analog value and the digital representation is known as quantization error.

Copying: Because of the inevitable presence of noise, making many successive copies of an analog communication is infeasible because each generation increases the noise. Because digital communications are generally error-free, copies of copies can be made indefinitely.

1.2 Post–turing Machine

Although digital signals are generally associated with the binary electronic digital systems used in modern electronics and computing, digital systems are actually ancient, and need be neither binary nor electronic.

A beacon is perhaps the simplest non-electronic digital signal, with just two states (on and off). In particular, smoke signals are one of the oldest examples of a digital signal, where an analog "carrier" (smoke) is modulated with a blanket to generate a digital signal (puffs) that conveys information.

More recently invented, a modem modulates an analog "carrier" signal (such as sound) to encode binary electrical digital information, as a series of binary digital sound pulses. A slightly earlier, surprisingly reliable version of the same concept was to bundle a sequence of audio digital "signal" and "no signal" information (i.e. "sound" and "silence") on magnetic cassette tape for use with early home computers.

In 1936 Alan Mathison Turing gave his answer to the question "What is a computable number?" by constructing his now well-known Turing machines as formalizations of the actions of a human computer. Less well-known is the almost synchronously published result by Emil Leon Post, in which a quasi-identical mechanism was developed for similar purposes.

A post-turing machine uses a binary alphabet, an infinite sequence of binary storage locations, and a primitive programming language with instructions for bi-directional movement among the storage locations and alteration of their contents one at a time.

The instructions may require the worker to perform the following "basic acts" or "operations":

- Marking the box he is in (assumed empty);
- Erasing the mark in the box he is in (assumed marked);
- Moving to the box on his right;

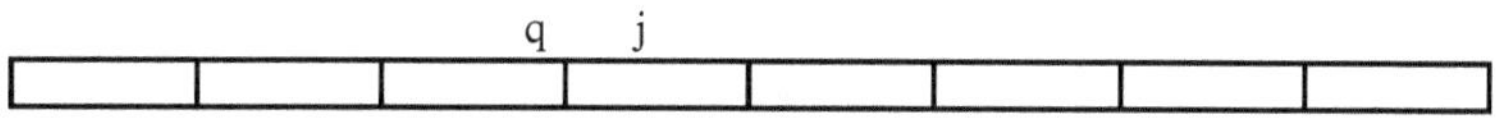

Fig. 1.3 Machine E.L. Post

- Moving to the box on his left;
- Determining whether the box he is in, is or is not marked.

In the hypothetical machine E.L. Post, information is represented in a binary alphabet $A = \{0,1\}$. The machine has an informational tape of unlimited length – the machine memory. Each cell can hold **0** or **1**. The machine has a "read head" (special sensor), which examines the contents of the cell (j), Fig. 1.3.

An informational tape can move in both directions, so that each move places the head in front of a particular cell.

The machine has a control unit, which at any one time is in a particular state – q. Tape is moved discretely so the head would stop in front of the cell.

Instruction set of the abstract machine:

1. Head to move to the right;
2. Head to move to the left;
3. Record label;
4. To erase the label;
5. To transfer control;
6. Stop.

For example, because the program looks for the hypothetical machine E.L. Post:

Move the control unit	Command number, i	Reference number, u	
right, 1 step	1	3	run i No 3
right, 1 step	2	4	run i No 4
Record label	3	2	run i No 2
The command transfer control	4	5	
Stop			

Each command is executed in one step, after which the command whose number is indicated in the u.

The Turing Machine differs from the Post machine in that the alphabet may have more than two characters, Fig. 1.4.

Each square of the tape holds exactly one of the symbols, also called input symbols or machine characters. It is assumed that one of the input symbols is a special one, the blank, denoted by **B**.

At any moment of time, the machine, being in one of its states and looking at one of the input symbols in some square, may act or halt. The action means that, in the next moment of time, the machine erases the old input symbol and writes a new input symbol on the same square (it may be the same symbol as before, or a new symbol; if the old one was not **B** and the new one is **B**, the machine is said to erase

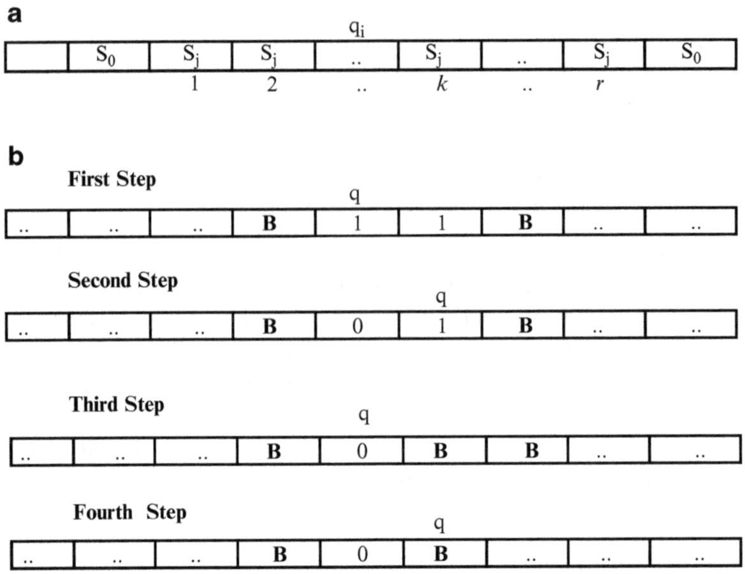

Fig. 1.4 The Turing Machine

Table 1.2 Turing machine program

Q \ A	0	1	*	s_0
q_2	–	q_2 0 L	q_1 * L	Q_0 S
q_1	q_2 1L	–	–	–
q_0	Stop			

the old symbol), changes the state to a new one (again, it is possible that the new state will be equal to the old one), and finally moves the head one square to the left, or one square to the right, or stays on the same square as before.

where:

S_0 – an empty cell;
S_{j1} – state (content) of the first non-empty left cell;
S_{jk} – state of the cell, which was seen at a given time;
r – The number of occupied cells;
q_i – state control device, $i=0,1,\ldots,m$.

A Turing machine program can be defined as a table (Table 1.2).

Assume that the initial configuration of the machine has the form: $s_0\ q_2\ 1 * 0\ s_0$. Then

$q_1\ 1 \rightarrow q_2\ 0$ Л	$s_0\ 0\ q_2 * 0\ s_0$	
$q_2\ * \rightarrow q_1\ *$ Л	$s_0\ 0 * q_1\ 0\ s_0$	
$q_1\ 0 \rightarrow q_1\ 1$ Л	$s_0\ 0 * 1\ q_2\ s_0$	
$q_2\ s_0 \rightarrow q_0\ s_0$ C	$s_0\ 0 * 1\ q_0\ s_0$	The final configuration of the machine

For some pairs of states and input symbols the action is not specified in the description of a Turing machine; thus the machine halts. In this case, symbols remaining on the tape form the output, corresponding to the original input, or more precisely, to the input string (or sequence) of input symbols. A sequence of actions, followed by a halt, is called a computation. A Turing machine accepts some input string if it halts on it. The set of all accepted strings over all the input symbols is called a language accepted by the Turing machine. Such languages are called recursively enumerable sets.

Another automaton is a nondeterministic Turing machine. It differs from an ordinary, deterministic Turing machine in that for a given state and input symbol, the machine has a finite number of choices for the next move. Each choice means a new input symbol, a new state, and a new direction to move its head.

A linear bounded automaton is a nondeterministic Turing machine which is restricted to the portion of the tape containing the input. The capability of the linear bounded automaton is smaller than that of a Turing machine.

Chapter 2
Number Systems, Operations, and Codes

Abstract In this chapter various number systems and conversions between them are presented and explained. The number systems concerned here are: the decimal system, the binary system, the hexadecimal system and the binary-coded decimal (BCD) code (i.e. an encoding for decimal numbers, in which each digit is represented by its own binary sequence to allow easier conversion to decimal digits and faster decimal calculations). The chapter, besides providing examples explaining conversions of whole numbers also shows the way of converting binary fractions to decimal ones (and opposite).

2.1 Number Systems

Convenient as the decimal number system generally is, its usefulness in machine computation is limited because of the nature of practical electronic devices. In most present digital machines, the numbers are represented, and the arithmetic operations performed, in a different number system called the binary number system. This chapter will help you more easily understand the structure of the binary number system, which is important in computers and digital electronics.

2.1.1 The Decimal System

In everyday life we use a system based on decimal digits (0, 1, 2, 3, 4, 5, 6, 7, 8, 9) to represent numbers and refer to the system as the decimal system. Consider what the number 62 means. It means six tens plus two:

$$62 = (6 \cdot 10) + 2.$$

The number 2678 means two thousands, six hundreds, seven tens, plus eight:

$$2678 = (2 \cdot 1000) + (6 \cdot 100) + (7 \cdot 10) + 8.$$

B. Borowik et al., *Theory of Digital Automata*, Intelligent Systems, Control and Automation: Science and Engineering 63, DOI 10.1007/978-94-007-5228-3_2,

The decimal system is said to have a **base**, or **radix**, of 10. This means that each digit in the number is multiplied by 10 raised to a power corresponding to that digit's position:

$$62 = \left(6 \cdot 10^1\right) + \left(2 \cdot 10^0\right);$$

$$2678 = \left(2 \cdot 10^3\right) + \left(6 \cdot 10^2\right) + \left(7 \cdot 10^1\right) + \left(8 \cdot 10^0\right)$$

Example: $A = 123{,}45$.

$$A = 1 \cdot 10^2 + 2 \cdot 10^1 + 3 \cdot 10^0 + 4 \cdot 10^{-1} + 5 \cdot 10^{-2},$$

$$A = 1964{,}52_{10} = 1 \cdot 10^3 + 9 \cdot 10^2 + 6 \cdot 10^1 + 4 \cdot 10^0 + 5 \cdot 10^{-1} + 2 \cdot 10^{-2}.$$

In general, for the decimal representation of $A = \{\ldots d_2 d_1 d_0 \cdot d_{-1} d_{-2} d_{-3} \ldots\}$, the value of A is

$$A = \sum_i d_i \cdot 10^i.$$

2.1.2 *The Binary System*

In the decimal system, 10 different digits are used to represent numbers with a base of 10. In the binary system, we have only two digits, 1 and 0. Thus, numbers in the binary system are represented to the base 2.

The binary numeral system, or base-2 number system, represents numeric values using two symbols, 0 and 1. Owing to its straightforward implementation in digital electronic circuitry using logic gates, the binary system is used internally by all modern computers.

The digits 1 and 0 in binary notation have the same meaning as in decimal notation:

$$0_2 = 0_{10}$$
$$1_2 = 1_{10}$$

To represent larger numbers, as with decimal notation, each digit in a binary number has a value depending on its position:

$$10_2 = \left(1 \cdot 2^1\right) + \left(0 \cdot 2^0\right) = 2_{10};$$

$$11_2 = \left(1 \cdot 2^1\right) + \left(0 \cdot 2^0\right) = 3_{10};$$

$$100_2 = \left(1 \cdot 2^2\right) + \left(0 \cdot 2^1\right) + \left(0 \cdot 2^1\right) = 4_{10},$$

Table 2.1 Decimal Binary and Hexadecimal numbers system

Decimal	Binary	Hexadecimal	Decimal	Binary	Hexadecimal	Decimal	Binary	Hexadecimal
0	0	0	11	1011	B	22	10110	16
1	1	1	12	1100	C	23	10111	17
2	10	2	13	1101	D	24	11000	18
3	11	3	14	1110	E	25	11001	19
4	100	4	15	1111	F	26	11010	1A
5	101	5	16	10000	10	27	11011	1B
6	110	6	17	10001	11	28	11100	1C
7	111	7	18	10010	12	29	11101	1D
8	1000	8	19	10011	13	30	11110	1E
9	1001	9	20	10100	14	31	11111	1F
10	1010	A	21	10101	15	32	100000	20

and so on. Again, fractional values are represented with negative powers of the radix:

$$1001{,}1101_2 = 1\cdot 2^3 + 0\cdot 2^2 + 0\cdot 2^1 + 1\cdot 2^0 + 1\cdot 2^{-1} + 1\cdot 2^{-2} + 0\cdot 2^{-3} + 1\cdot 2^{-4}.$$

In general, for the binary representation of $B = \{\ldots b_2 b_1 b_0 \cdot b_{-1} b_{-2} b_{-3} \ldots\}$, the value of B is

$$B = \sum_i b_i \cdot 2^i.$$

2.1.3 *Hexadecimal Notation*

Because of the inherent binary nature of digital computer components, all forms of data within computers are represented by various binary codes. However, no matter how convenient the binary system is for computers, it is exceedingly cumbersome for human beings. Consequently, most computer professionals who must spend time working with the actual raw data in the computer prefer a more compact notation.

What notation to use? One possibility is the decimal notation. This is certainly more compact than binary notation, but it is awkward because of the tediousness of converting between base 2 and base 10.

Instead, a notation known as hexadecimal has been adopted. Binary digits are grouped into sets of four. Each possible combination of four binary digits is given a symbol, as follows (Table 2.1):

Because 16 symbols are used, the notation is called **hexadecimal**, and the 16 symbols are the hexadecimal digits.

A sequence of hexadecimal digits can be thought of as representing an integer in base 16.

Thus,

$$2C_{16} = \left(2_{16}\cdot 16^1\right) + \left(C_{16}\cdot 16^0\right) = \left(2_{10}\cdot 16^1\right) + \left(2_{10}\cdot 16^0\right) = 44.$$

Table 2.2 ANCII, Decimal, Binary and Hexadecimal code

Symbols ANCII	Decimal code	Binary code	Hexadecimal code
0	48	0110000	30
1	49	0110001	31
2	50	0110010	32
A	65	1000001	41
B	66	1000010	42
F	70	1000110	46
:	58	0111010	3F
(	40	0101000	28

Each hexadecimal digit represents four binary digits (bits). For example, byte values can range from 0 to 255 (decimal) but may be more conveniently represented as two hexadecimal digits in the range 00 through FF. Hexadecimal is also commonly used to represent computer memory addresses.

Hexadecimal notation is used not only for representing integers. It is also used as a concise notation for representing any sequence of binary digits, whether they represent text, numbers, or some other type of data, Table 2.1. The reasons for using hexadecimal notation are:

- It is more compact than binary notation.
- In most computers, binary data occupy some multiple of four bits, and hence some multiple of a single hexadecimal digit.
- It is extremely easy to convert between binary and hexadecimal (Table 2.2).

2.1.4 Binary-Coded Decimal Code

In computing and electronic systems, binary-coded decimal (BCD) or, in its most common modern implementation, packed decimal, is an encoding for decimal numbers in which each digit is represented by its own binary sequence. Its main virtue is that it allows easy conversion to decimal digits for printing or display, and allows faster decimal calculations. Its drawbacks are a small increase in the complexity of circuits needed to implement mathematical operations. Uncompressed BCD is also a relatively inefficient encoding -it occupies more space than a purely binary representation.

In BCD, a digit is usually represented by four bits which, in general, represent the decimal digits 0 through 9. Other bit combinations are sometimes used for a sign or for other indications (e.g., error or overflow). To encode a decimal number using the common BCD encoding, each decimal digit is stored in a 4-bit nibble, Table 2.3.

Table 2.3 Examples of binary-coded decimal code

Decimal	BCD	Decimal	BCD	Decimal	BCD
0	0000	10	0001 0000	20	0010 0000
1	0001	11	0001 0001	21	0010 0001
2	0010	12	0001 0010	22	0010 0010
3	0011	13	0001 0011	33	0011 0011
4	0100	14	0001 0100	34	0011 0100
5	0101	15	0001 0101	45	0100 0101
6	0110	16	0001 0110	56	0101 0110
7	0111	17	0001 0111	67	0110 0111
8	1000	18	0001 1000	78	0111 1000
9	1001	19	0001 1001	89	1000 1001

Although uncompressed BCD is not as widely used as it once was, decimal fixed-point and floating-point are still important and continue to be used in financial, commercial, and industrial computing.

2.2 Converting Between Number Systems

2.2.1 Converting Between Binary and Decimal (Between Decimal and Binary)

It is a simple matter to convert a number from binary notation to decimal notation. In fact, we showed several examples in the previous subsection. All that is required is to multiply each binary digit by the appropriate power of 2 and add the results.

To convert from decimal to binary, the integer and fractional parts are handled separately.

2.2.2 Integers

For the integer part, recall that in binary notation, an integer is represented by

$$b_{m-1}b_{m-2}\dots b_2 b_1 b_0 \quad b_i = 0 \quad \text{or} \quad 1.$$

Suppose it is required to convert a decimal integer A into binary form. If we divide A by 2, in the decimal system, and obtain a quotient A_1 and a remainder R_0, we may write

$$A = 2 \cdot A_1 + R_0 \quad R_0 = 0 \quad \text{or} \quad 1.$$

Next, we divide the quotient A_1 by 2. Assume that the new quotient is A_2 and the new remainder R_1. Then

$$A_1 = 2 \cdot A_2 + R_1 \quad R_1 = 0 \quad \text{or} \quad 1$$

so that

$$A = 2 \cdot (2 \cdot A_2 + R_1) + R_0 = (A_2 \cdot 2^2) + (R_1 \cdot 2^1) + R_0.$$

If next,

$$A_2 = 2 \cdot A_3 + R_2.$$

Because $A > A_1 > A_2 > \cdots$, continuing this sequence will eventually produce a quotient $A_{m-1} = 1$ (except for the decimal integers 0 and 1, whose binary equivalents are 0 and 1, respectively) and a remainder R_{m-2}, which is 0 or 1. Then

$$A = (1 \cdot 2^{m-1}) + (R_{m-2} \cdot 2^{m-2}) + \cdots + (R_2 \cdot 2^2) + (R_1 \cdot 2^1) + R_0$$

which is the binary form of A. Hence, we convert from base 10 to base 2 by repeated divisions by 2. The remainders and the final quotient, 1, give us, in order of increasing significance, the binary digits of A.

Example. Example of Converting from Decimal Notation to Binary Notation for Integers

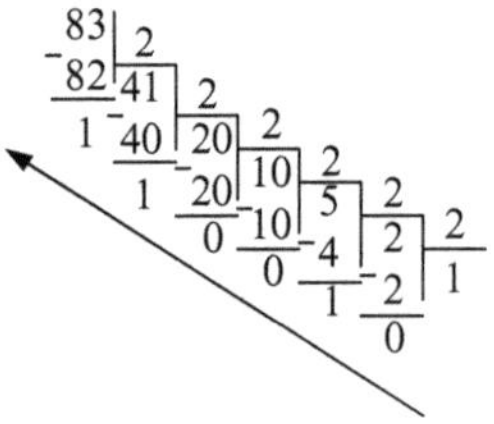

Thus, we obtain $83_{10} = 1010011_2$.

2.2.3 Fractions

For the fractional part, recall that in binary notation, a number with a value between 0 and 1 is represented by

$$0.b_{-1}b_{-2}b_{-3} \ldots b_{-m} \quad b_i = 0 \quad or \quad 1$$

and has the value

$$\left(b_{-1}\cdot 2^{-1}\right)+\left(b_{-2}\cdot 2^{-2}\right)+\left(b_{-3}\cdot 2^{-3}\right)+\ldots+\left(b_{-\mathrm{m}}\cdot 2^{-\mathrm{m}}\right)$$

This can be rewritten as

$$2^{-1}\cdot(b_{-1}+2^{-1}\cdot(b_{-2}+2^{-1}\cdot(b_{-3}+2^{-1}\cdot(\ldots+2^{-1}\cdot(b_{-\mathrm{m}+1}+2^{-1}\cdot b_{-\mathrm{m}})\ldots))))$$

This expression suggests a technique for conversion. Suppose we want to convert the number F $(0<F<1)$ from decimal to binary notation. We know that F can be expressed in the form

$$F=2^{-1}\cdot(b_{-1}+2^{-1}\cdot(b_{-2}+2^{-1}\cdot(b_{-3}+2^{-1}\cdot(\ldots+2^{-1}\cdot(b_{-m+1}+2^{-1}b_{-m})\ldots))))$$

We can say that $\left(2\cdot F\right)=b_{-1}+F_1$, where $0<F_1<1$ and where

$$F_1=2^{-1}\cdot(b_{-2}+2^{-1}\cdot(b_{-3}+2^{-1}\cdot(\ldots+2^{-1}\cdot(b_{-\mathrm{m}+1}+2^{-1}b_{-\mathrm{m}})\ldots)))$$

To find b_{-2}, we repeat the process. Therefore, the conversion algorithm involves repeated multiplication by 2. At each step, the fractional part of the number from the previous step is multiplied by 2. The digit to the left of the decimal point in the product will be 0 or 1 and contributes to the binary representation, starting with the most significant digit. The fractional part of the product is used as the multiplicand in the next step.

Example. Example of converting from decimal notation to binary notation for fractions

$$
\begin{array}{r}
{}_{\times}0{,}3125\\
2\\ \hline
{}_{\times}0{,}6250\\
2\\ \hline
{}_{\times}1{,}2500\\
2\\ \hline
{}_{\times}0{,}5000\\
2\\ \hline
1{,}0000
\end{array}
\downarrow
$$

Thus, we obtain $0{,}3125_{10}=0{,}0101_2$.

This process is not necessarily exact; that is, a decimal fraction with a finite number of digits may require a binary fraction with an infinite number of digits. In such cases, the conversion algorithm is usually halted after a prespecified number of steps, depending on the desired accuracy.

Example. Convert number $118,\ 376_{10}$ from decimal code in binary code.

2.2.4 Integers

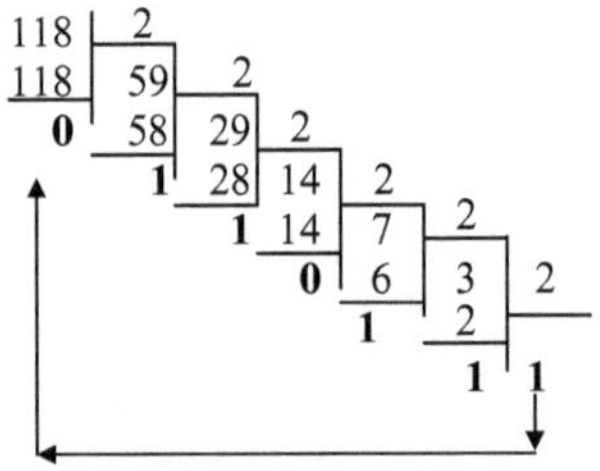

Thus, we obtain $118_{10} = 1110110_2$.

2.2.5 Fractions

$$\begin{array}{l} 0,\ 376_{10} \\ \underline{\times \qquad\ \ 2} \\ 0,\ 752_{10} \\ \underline{\times \qquad\ \ 2} \\ 1,\ 504_{10} \\ \underline{\times \qquad\ \ 2} \\ 1,\ 008_{10} \end{array}$$

Thus, we obtain $0{,}376_{10} \approx 0{,}011_2$.
The final result $118{,}\ 376_{10} \approx 1110110{,}011_2$.

Example. Example of converting from binary notation to decimal notation for integers

$$100111 = 1\cdot 2^5 + 0\cdot 2^4 + 0\cdot 2^3 + 1\cdot 2^2 + 1\cdot 2^1 + 1\cdot 2^0 = 32 + 4 + 2 + 1 = 39.$$

Example. Example of converting from binary notation to decimal notation

$$\begin{aligned} 1011{,}01101 &= 1\cdot 2^3 + 0\cdot 2^2 + 1\cdot 2^1 + 1\cdot 2^0 + 0\cdot 2^{-1} + 1\cdot 2^{-2} + 1\cdot 2^{-3} + 0\cdot 2^{-4} + 1\cdot 2^{-5} \\ &= 8 + 2 + 1 + \frac{1}{4} + \frac{1}{8} + \frac{1}{32} = 11\frac{13}{32}. \end{aligned}$$

The decimal value of any binary number can be found by adding the weights of all bits that are 1 and discarding the weights of all bits that are 0.

Example. Convert the binary whole number 10111010 to decimal.

In order to represent the 10 decimal digits 0, 1,…, 9, it is necessary to use at least 4 binary digits. Since there are 16 combinations of 4 binary digits, of which 10 combinations are used, it is possible to form a very large number of distinct codes.

Table 2.4

Weight:	1	2	3	4	5	6	7	8	9
1	1	2	3	4	5	6	7	8	9
10	A	14	1E	28	32	3C	46	50	5A
10^2	64	C8	12C	190	1F4	258	2BC	320	384
10^3	3E8	7D0	BB8	FA0	1388	1770	1B58	1F40	2328
10^4	2710	4E20	7530	9C40	C350	EA60	11170	13880	15F90
10^5	186A0	30D40	493E0	61A80	7A120	927C0	AAE60	C3500	DBBA0

Example: $1234 = 1000 + 200 + 30 + 4 = (3E8)_{16} + (C8)_{16} + (1E)_{16} + (4)_{16} = (4D2)_{16}$

Of particular importance is the class of weighted codes, whose main characteristic is that each binary digit is assigned a decimal "weight," and, for each group of four bits, the sum of the weights of those binary digits whose value is 1 is equal to the decimal digit which they represent.

Binary number:	1	0	1	1	1	0	1	0
Weight:	2^7	2^6	2^5	2^4	2^3	2^2	2^1	2^0
Result	128	0	32	16	8	0		

Thus, we obtain $128 + 0 + 32 + 16 + 8 + 0 + 2 + 0 = 186$.

Result – $10111010_2 = 186_{10}$.

Converting between Decimal (Binary) code and hexadecimal code (Table 2.4)

Example. Hexadecimal number $9F2_{16}$ converted in Binary code:

9	F	2
↓	↓	↓
1001	1111	0010

Thus, we obtain $9F2_{16} = 100111110010_2$.

Example. Hexadecimal number $IFA,C24_{16}$ converted in Binary code:

I	**F**	**A,**	**C**	**2**	**4**
↓	↓	↓	↓	↓	↓
0001	**1111**	**1010,**	**1100**	**0010**	**0100**

Thus, we obtain $IFA,C24_{16} = 111111010,110000100100_2$.

Example. Binary number 10110110001101_2 converted in Hexadecimal code:

$$10110110001101_2 = \underline{0010}\ \underline{1101}\ \underline{1000}\ \underline{1101} = 2D8D_{16}.$$

2.2.6 *Binary to BCD Conversion*

The basic idea is to shift data serially into a shift register. As each bit is shifted in, the accumulated sum is collected. Each shift effectively doubles the value of the binary number in the four-bit shift register which is going to hold the converted BCD digit.

Each time a bit is shifted in, the value in the shift register is doubled. After 4 bits have been shifted in, if the original value is 0, 1, 2, 3, or 4, then the result is within the 0–9 range of a BCD digit and there is no action required.

If the value is 5, 6, 7, 8, or 9, then the doubled result is greater than 10, so a carry out (called ModOut in the code) is generated to represent the overflow into the tens column (i.e. into the next BCD digit).

Example. $A=49_{10}=110001_2$.

									110001
(1) Shift								1	10001
(2) Shift							1	1	0001
(3) Shift						1	1	0	001
(4) Shift					1	1	0	0	01
and correction									
+0110					0	1	1	0	
Result ·				1	0	0	1	0	01
(5) Shift			1	0	0	1	0	0	1
(6) Shift		1	0	0	1	0	0	1	
Result	**0**	**1**	**0**	**0**	**1**	**0**	**0**	**1**	

The material on number systems is available in almost all elementary texts on algebra, switching theory, and digital computers. An extensive discussion of computer arithmetic is available in [5–11].

Chapter 3
Binary Arithmetic

Abstract This chapter introduces to the methods for adding and multiplying binary numbers. The topic is developed by first considering the binary representation of unsigned numbers (which are the easiest to understand), and then binary representation of signed numbers and fractions (the hardest to understand) are concerned. Binary arithmetic is essential in all digital computers and in many other types of digital systems. To understand digital systems, the basics of binary addition, subtraction, multiplication, and division must be known.

3.1 Binary Addition

Adding binary numbers is a very simple task, and very similar to the longhand addition of decimal numbers. As with decimal numbers, you start by adding the bits (digits) one column, or place weight, at a time, from right to left.

(a)	(b)	(c)
0.0010 (2_{10})	1010 (10_{10})	0.100111 (39_{10})
+0.0100 (4_{10})	+1011 (11_{10})	±0.001101 (13_{10})
0.0110 (6_{10})	10101 (21_{10})	0.110100 (52_{10})

Notice that the first three rules result in a single bit and in the fourth rule the addition of two 1s yields a binary two (10). When binary numbers are added, the last condition creates 10 a sum of 0 in a given column and a carry of 1 over to the next column to the left, as illustrated in the following example:

	Decimal code	Binary code
Carry bits	166	1 111
	47	10100110
		00101111
Result	213	11010101

B. Borowik et al., *Theory of Digital Automata*, Intelligent Systems, Control and Automation: Science and Engineering 63, DOI 10.1007/978-94-007-5228-3_3,

Table 3.1 Binary arithmetic

The four basic rules for adding binary digits (bits) are as follows:	The four basic rules for subtracting bits are as follows:	The four basic rules for multiplying bits are as follows:
0+0=0	0−0=0	$0 \times 0=0$
1+0=1	1−0=1	$1 \times 0=0$
0+1=1	1−1=0	$0 \times 1=0$
1+1=10	10−1=1	$1 \times 1=1$

3.2 Binary Subtraction

Subtraction is generally simpler than addition since only two numbers are involved and the upper value representation is greater than the lower value representation. The problem of "borrow" is similar in binary subtraction to that in decimal. We can construct a subtraction table that has two parts – the three cases of subtracting without borrow, and the one case of the involvement of a borrow digit, no matter how far to the left is the next available binary digit.

(a)	(b)
10101 (21_{10})	1000 (8_{10})
−1010 (10_{10})	−11 (3_{10})
1011 (11_{10})	101 (5_{10})

When subtracting numbers, you sometimes have to borrow from the next column to the left. A borrow is required in binary only when you try to subtract a 1 from a 0. In this case, when a 1 is borrowed from the next column to the left, a 10 is created in the column being subtracted, and the last of the four basic rules just listed must be applied, Table 3.1.

For 10 minus 1, 1 is borrowed from the "tens" column for use in the "ones" column, leaving the "tens" column with only 2. The following examples show "borrowing" in binary subtraction.

(c)	(d)	(e)
0.0010 (2_{10})	100 (4_{10})	0.1010 (10_{10})
−0.0001 (1_{10})	−010 (2_{10})	−0.0110 (6_{10})
0.0001 (1_{10})	010 (2_{10})	0.0100 (4_{10})

3.3 Binary Multiplication

Binary multiplication of two bits is the same as multiplication of the decimal digits 0 and 1. Multiplication is performed with binary numbers in the same manner as with decimal numbers. It involves forming partial products, shifting each successive partial product left one place, and then adding all the partial products.

0.0101 (5_{10})	1101	1101
× 0.0011 (3_{10})	× 1101	× 1101
0101	1101	1101
+ 0101	+ 0000	+ 1101
0.1111 (15_{10})	1101	0000
	1101	1101
	10101001	**10101001**
(a)	(b)	(c)

3.4 Binary Division

Basically the reverse of the multiply by shift and add. Division in binary code follows the same procedure as division in decimal code.

$A = 430_{10} = 110101110_2$;	$A = 204_{10} = 11001100_{(2)}$,
$B = 10_{10} = 1010$; $A/B = 43_{10}$	$B = 12_{10} = 1100_{(2)}$,
	$/B = 204_{10}/12_{10} = 17_{10}$
110101110 | 1010	11001100 | 1100
−1010 101011	1100 | 10001
1101	00001
− 1010	− 0
1111	11
− 1010	− 0
1010	110
− 1010	− 0
0000	1100
	− 1100
	0000
(a)	(b)

Detailed study of digital arithmetic is beyond the scope of this book. For a more comprehensive discussion of computer arithmetic, the reader may consult [4, 5, 10, 11].

3.5 BCD Addition

The procedures followed in adding BCD are the same as those used in binary. For example, let's consider the addition of the two BCD digits 5 and 3:

+0101	(BCD)	(5_{10})
0011	(BCD)	(3_{10})
1000	(BCD)	(8_{10})

There is, however, the possibility that addition of BCD values will result in invalid totals. The following example shows this:

	+1001	(BCD)	(9_{10})
	0110	(BCD)	(6_{10})
Invalid BCD →	1111		(15_{10})

The sum 1111_2 is the binary equivalent of 15_{10}; however, 1111 is not a valid BCD number. You cannot exceed 1001 in BCD, so a correction factor must be made. To do this, you add 6_{10} (0110_{BCD}) to the sum of the two numbers. The "**add 6**" correction factor is added to any BCD group larger than 1001_2.

Remember, there is no 1010_2, 1011_2, 1100_2, 1101_2, 1110_2, or 1111_2 in BCD:

+ 1111		← Invalid BCD
0110	(BCD)	Add (6_{10})
0001 1111		← New BCD

The sum plus the add 6 correction factor can then be converted back to decimal to check the answer.

Add two numbers $A = 279_{10} = 0010\ 0111\ 1001$, $B = 581_{10} = 0101\ 1000\ 0001$.

0010	0111	1001	
+0101	1000	0001	
0111	1111	1010	
+	0110	0110	Add (6_{10})
C = 1000 ←	0110 ←	0000	

Result $C = 100001100000 = 860_{10}$.

3.6 Arithmetic Operations with Signed Numbers

The one's and two's complements of a binary number are operations used by computers, to perform internal mathematical calculations. To complement a binary number means to change it to a negative number.

3.6.1 1s and 2s Complements Forms

This allows the basic arithmetic operations of subtraction, multiplication, and division to be performed through successive addition. The intention of this section is to introduce the basic concepts of complementing.

3.6.2 1s Complement

Let's assume that we have a 5-bit binary number that we wish to represent as a negative number. The number is decimal 19, or binary:

$$10011_2$$

There are two ways to represent this number as a negative number. The first method is to simply place a minus sign in front of the number, as we do with decimal numbers:

$$-(10011)_2$$

This method is suitable for us, but it is impossible for computers to interpret, since the only symbols they use are binary 1s and 0s. To represent negative numbers, then, some digital computing devices use what is known as the one's complement method. First, the one's complement method places an extra bit (sign bit) in the most significant (left-most) position and lets this bit determine whether the number is positive or negative. The number is positive if the sign bit is 0 and negative if the sign bit is 1. Using the one's complement method, +19 decimal is represented in binary as shown here with the sign bit (0) indicated in bold:

$$\mathbf{0}\ 10011_2$$

The negative representation of binary 10011 is obtained by placing a 1 in the most significant bit position and inverting each bit in the number (changing 1s to 0s and 0s to 1s). So, the one's complement of binary 10011 is:

$$\mathbf{1}\ 01100_2$$

If a negative number is given in binary, its one's complement is obtained in the same fashion.

$$+15_{10}=\mathbf{0}\ 1111_2$$
$$-15_{10}=\mathbf{1}\ 0000_2$$

3.6.3 2s Complement

The two's complement is similar to the one's complement in the sense that one extra digit is used to represent the sign. The two's complement computation, however, is slightly different. In the one's complement, all bits are inverted; but in the two's complement, each bit, from right to left, is inverted only after the first 1 is detected. Let's use the number +22 decimal as an example:

$$+22_{10}=\mathbf{0}\ 10110_2$$

Its two's complement would be:

$$-22_{10}=\mathbf{1}\ 01010_2$$

Note that in the negative representation of the number 22, starting from the right, the first digit is a 0, so it is not inverted; the second digit is a 1, so all digits after this one are inverted.

Table 3.2 4-bit 1s and 2s complements forms

	1s and 2s complements forms				1s and 2s complements forms		
Decimal	Binary	4-bit 1s complement	4-bit 2s complement	Decimal	Binary	4-bit 1s complement	4-bit 2s complement
+7	0.111	0.111	0.111	−0	1.000	1.111	0.000
+6	0.110	0.110	0.110	−1	1.001	1.110	1.111
+5	0.101	0.101	0.101	−2	1.010	1.101	1.110
+4	0.100	0.100	0.100	−3	1.011	1.100	1.101
+3	0.011	0.011	0.011	−4	1.100	1.011	1.100
+2	0.010	0.010	0.010	−5	1.101	1.010	1.011
+1	0.001	0.001	0.001	−6	1.110	1.001	1.010
+0	0.000	0.000	0.000	−7	1.111	1.000	1.001

Table 3.3 8-bit 1s and 2s complements forms

Decimal	Binary	Inverse (1s complement)	2s complement
−7	$[-7]_b=1.000111_2$	$[-7]_i=1.111000_2$	$[-7]_{tc}=1.111001_2$

If a negative number is given in two's complement, its complement (a positive number) is found in the same fashion:

$$-14_{10}=\mathbf{1}\ 10010_2$$
$$+14_{10}=\mathbf{0}\ 01110_2$$

Again, all bits from right to left are inverted after the first 1 is detected. Other examples of the 1s and 2s complement are shown here (Tables 3.2 and 3.3):

3.6.4 Additional in the 1s Complement System

There are several options for adding:

Case 1. $A>0, B>0, A+B<1.\ [A>0]_i+[B>0]_i=A+B.$

Case 2. $A>0, B<0, A+B>0.\ [A>0]_i+[B<0]_i=A+2+B-2^{-n}$

Case 3. $A>0, B<0, A+B<0.\ [A>0]_i+[B<0]_i=A+2+B-2^{-n}$

Case 4. $A<0, B<0, |A+B|<1.\ [A<0]_i+[B<0]_i=2+A-2^{-n}+2+B-2^{-n}$

Where the $[A]_i$, $[B0]_i$ – representation of numbers in a computer.

Examples

Case 2.

$A=+0{,}1101 \quad [A]_b=0{,}1101 \quad [A]_i=0{,}1101$

$B=-0{,}0011 \quad [B]_b=1{,}0011 \quad [B]_i=1{,}1100$

$\mathbf{1}\leftarrow 0{,}1001$

1

$C=0{,}1010 \Leftarrow [C]_b=0{,}1010 \Leftarrow [C]_i=0{,}1010$

Case 3.

$A=-0{,}1101 \quad [A]_b=1{,}1101 \quad [A]_i=1{,}0010$

$B=+0{,}0011 \quad [B]_b=0{,}0011 \quad \underline{[B]_i=0{,}0011}$

$C=-0{,}1010 <= [C]_b= 1{,}1010 <= [C]_i=1{,}0101$

Case 4.

$A=-0{,}0101 \quad [A]_b=1{,}0101 \quad [A]_i=1{,}1010$

$B=-0{,}0110 \quad [B]_b=1{,}0110 \quad \underline{[B]_i=1{,}1001}$

$\mathbf{1}\leftarrow 1{,}0011$

$\underline{\qquad\quad 1}$

$C=-0{,}1011 <= [C]_b=1{,}1011 <= [C]_i=1{,}0100$

Example. Case $|A|=|B|,\ A<0,\ B>0.$

$A=-0{,}0101 \quad [A]_{db}=1{,}0101 \quad [A]_i=1{,}1010$

$B=+0{,}0101 \quad [B]_b=0{,}0101 \quad \underline{[B]_i=0{,}0101}$

$C=-0{,}0000 <= [C]_b=1{,}0000 <= [C]_i=1{,}1111$

Example. Case $A>B,\ B>0,\ |A+B|=1.$

$A=+0{,}0111 \quad [A]_b=0{,}0111 \quad [A]_i=0{,}0111$

$B=+0{,}1001 \quad [B]_b=0{,}1001 \quad \underline{[B]_i=0{,}1001}$

$C=+0{,}1111 <= [C]_b=0{,}1111 <= [C]_i=1{,}0000$

Example. Case $A<0,\ B<0,\ |A+B|=1.$

$A=-0{,}0111 \quad [A]_b=1{,}0111 \quad [A]_i=1{,}1000$

$B=-0{,}1001 \quad [B]_b=1{,}1001 \quad \underline{[B]_i=1{,}0110}$

$\mathbf{1}\leftarrow 0{,}1110$

$\underline{\qquad\quad 1}$

$C=+0{,}1111 <= [C]_b=0{,}1111 <= [C]_i=0{,}1111$

3.6.5 Additional in the 2s Complement System

There are several options for adding:

Case 1. A>0, $B>0,\ A+B<1.\ [A>0]_{tc}+[B>0]_{tc}=A+B.$

Case 2. $A>0,\ B<0,\ A+B>0.\ [A>0]_{tc}+[B<0]_{tc}=A+2+B.$

Case 3. $A>0,\ B<0,\ A+B<0.\ [A>0]_{tc}+[B<0]_{tc}=A+2+B.$

Case 4. $A<0,\ B<0, |A+B|<1.\ [A<0]_{tc}+[B<0]_{tc}=2+A+2+B.$

Here the $[A]_{tc}, [B0]_{tc}$– representation of numbers in a computer.

Examples.

Case 3.

$A=-\ 0{,}1101\ [A]_b=1{,}1101\ [A]_{tc}=1{,}0011$

$B=+\ 0{,}0011\ [B]_b=0{,}0011\ \underline{[B]_{tc}=0{,}0011}$

$C=-0{,}1010 <= [C]_b=1{,}1010 <= [C]_{tc}=1{,}0110$

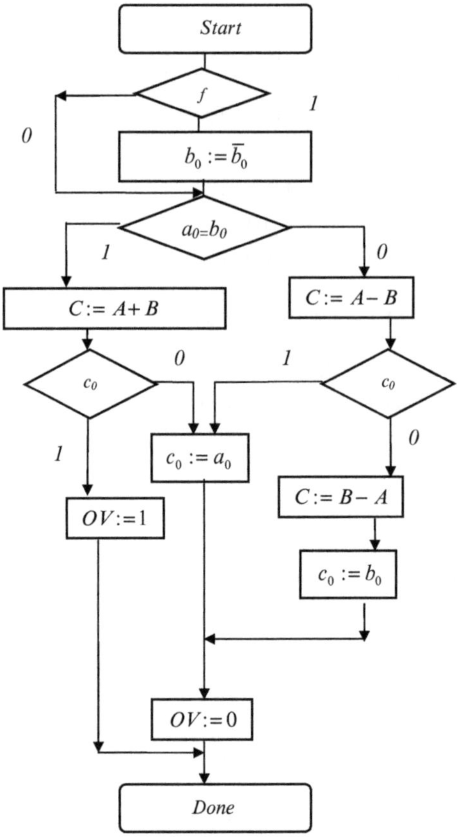

where a_o, b_0, c_o – sign bits;
OV - value of overflow digit;
f- The type of operation (*f=0 – Addition, f=1 – Subtraction*)

Fig. 3.1 The block diagram of the addition (subtraction) of binary numbers in the direct code

Case 4.

$A = -\,0{,}0101$ $[A]_b = 1{,}0101$ $[A]_{tc} = 1{,}1011$
$B = -\,0{,}0110$ $[B]_b = 1{,}0110$ $[B]_{tc} = 1{,}1010$
$C = -\,0{,}1011 <=$ $[C]_b = 1{,}1011 <=$ $[C]_{tc}^{-} = \mathbf{1}\ 1{,}0101$

Example. Case $|A| = |B|$, $A < 0$, $B > 0$.

$A = -0{,}0101$ $[A]_b = 1{,}0101$ $[A]_{tc} = 1{,}1011$
$B = +\,0{,}0101$ $[B]_b = 0{,}0101$ $[B]_{tc} = 0{,}0101$
$C = +0{,}0000 <=$ $[C]_b = 0{,}0000 <=$ $[C]_{tc}^{-} = \mathbf{1}\ 0{,}0000$

Figures 3.1, 3.2, 3.3 show a block diagram of the addition (subtraction) of binary numbers in the direct, inverse and complementary codes.

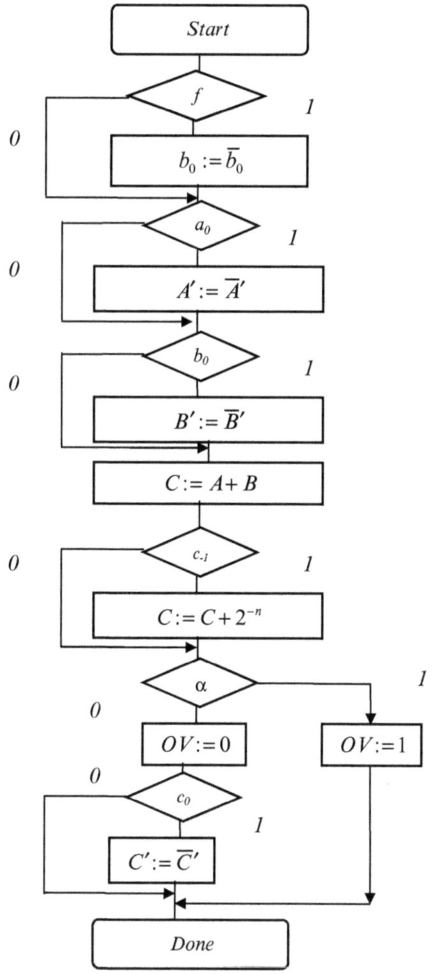

where a_o, b_0, c_o – sign bits;

OV - value of overflow digit;

A', B', C' -modules of numbers;

c_{-1} – transfer from a sign bit;

f- The type of operation (*f=0 – Addition, f=1 – Subtraction*)

Fig. 3.2 The block diagram of the addition (subtraction) of binary numbers in the inverse code (1s complement form)

3.7 BCD Subtraction

Either packed or unpacked BCD numbers can be subtracted. BCD subtraction follows the same rules as binary subtraction. However, if the subtraction causes a

Fig. 3.3 The block diagram of the addition (subtraction) of binary numbers in the 2s complement form

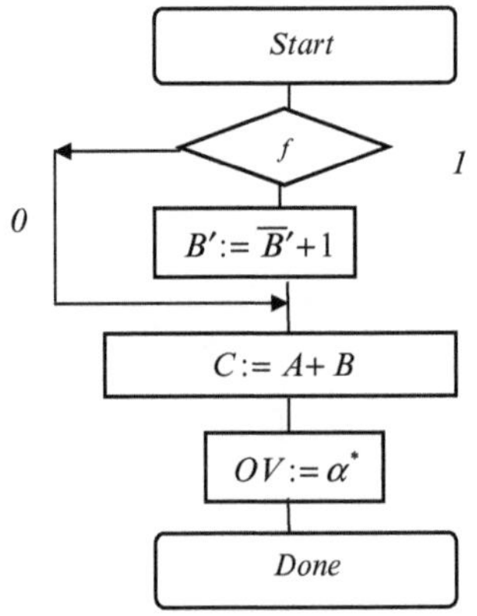

where a_o, b_0, c_o – sign bits;

OV - value of overflow digit;

α^* - value of overflow in an additional code;

B' - module of number;

f- The type of operation (*f=0 – Addition, f=1 – Subtraction*)

borrow and/or creates an invalid BCD number, an adjustment is required to correct the answer. The correction method is to subtract 6 from the difference in any digit position that has caused an error.

Examples.

Represent the number of $A = -256_{10}$ in the inverse code for the BCD:

$$\begin{array}{lrl} & 1.\,0010\ 0101\ 0110 & \\ + & \underline{0110\ 0110\ 0110} & \text{Add } (6_{10})\ 0110 \\ & 1000\ 1011\ 1100 & \\ \textbf{Result} & A_i = 1.\,0111\ 0100\ 0011. & \end{array}$$

Represent the number of $A = -398_{10}$ in the two complement code for the BCD:

$$\begin{array}{lrl} & 1.\,0011\ 1001\ 1000 & \text{Add } (6_{10})\ 0110 \\ + & \underline{0110\ 0110\ 0110} & \\ & 1.\,1001\ 1111\ 1110 & \\ & 0110\ 0000\ 0001 & \\ & \underline{+1} & \\ \textbf{Result } A_{tc} & = 1.0110\ 0000\ 0010 & \end{array}$$

$A = 37_{10} = 0011\ 0111_{(BCD)}$. $B = 12_{10} = 0000\ 0010_{(BCD)}$.

Result $C = A - B = 15_{10} =$ BCD ?

	0011 0111
+	0000 0010
	0011 0101
Result	$C=0011\ 0101_{(BCD)}=25_{10}$

$A=-1000\ 0010\ 0101_{(BCD)}=-825_{10}$, $B=1001\ 0100\ 0110_{(BCD)}=946_{10}$.
Result $C=-A+B=(-825+946)_{10}=121_{10}=$ BCD ?

	1. 0001 0111 0101
	0. 1001 0100 0110
	1. 1010 1011 1011
+	0110 0110 0110 (adjustment Add $0110_{(BCD)}$)
Result	$C=0. \leftarrow 0001 \leftarrow 0010 \leftarrow 0001.$

3.8 BCD Multiplication and Division

Multiplication cannot be performed on packed BCD; the four most significant bits must be zeroed for the adjustment to work.

BCD division also cannot be performed on packed numbers. Before dividing an unpacked BCD number, the division adjustment is made by converting the BCD numbers to binary.

Example. $A=25_{10}=0010\ 0101$ на $B=12_{10}=0001\ 0010$. Intermediate results of multiplication put in P. $A{\cdot}B=0010\ 0101 \cdot 0001\ 0010=0011\ 0000\ 0000=300_{10}$.

P 0000 0000 0000	
+A 0010 0101	$0010-0001=0001>0$, Repeat $B+P$
P 0000 0010 0101	
±A 0010 0101	$0010-0001=0$
P 0000 0100 1010	
+0110	adjustment
P 0000 0101 0000	shift B on the left 4 bits and add to P
P 0000 0101 0000	
+A 0010 0101 0000	$0001-0001=0$
P 0010 1010 0000	
+ 0110	adjustment
P 0011 0000 0000= $=300_{10}$	$A{\cdot}B=0011\ 0000\ 0000=300_{10}$

Example.

$A=48_{10}=0100\ 1000$, $B=2_{10}=0000\ 0010$, $A/B=24_{10}=0010\ 0100$. In the $C1$ – form the older BCD tetrad private, and $C2$ – the LSB.

Sign Bit	p	Sign Bit m	m (23 bit)
0 1	k	0 1	

Fig. 3.4 Floating-point arithmetic represented as integers ($N = m \cdot q^p$)

$A/B = 0100\ 1000/0010$	
$-\ 0010$	
$0010 > 0$	$C1 = C1 + 1 = 1$
$-\ 0010$	
0000	
0010	$C1 = 1 + 1 = 2 = 0010$
$-\ 0010$	
$\pm\ 0010$	
0000	*B shift to 4 digits to the right and perform the same steps:*
$0100\,1000$	
-0010	
$0110 > 0$	$C2 = C2 + 1 = 1$
-0010	
$0100 > 0$	$C2 = 1 + 1 = 2$
-0010	
$0010 > 0$	$C2 = 2 + 1 = 3$
-0010	
0000	$C2 = 3 + 1 = 4 = 0100$

Result $C1 + C2 = 0010\ 0000 + 0000\ 0100 = 0010\ 0100 = 24_{10}$

3.9 Floating-Point Numbers

In computing, floating-point describes a system for representing numbers that would be too large or too small to be represented as integers. Numbers ($N = m \cdot q^p$) are in general represented approximately to a fixed number of significant digits and scaled using an exponent. The base (q) for the scaling is normally 2, 10 or 16. The typical number that can be represented exactly is of the form, Fig. 3.4.

The term floating point refers to the fact that the radix point (decimal point, or, more commonly in computers, binary point) can "float"; that is, it can be placed anywhere relative to the significant digits of the number. This position is indicated separately in the internal representation, and floating-point representation can thus be thought of as a computer realization of scientific notation. Over the years, several different floating-point representations have been used in computers; however, for the last 10 years the most commonly encountered representation is that defined by the IEEE 754 Standard.

The JVM's floating-point support adheres to the IEEE-754 1985 floating-point standard. This standard defines the format of 32-bit and 64-bit floating-point numbers and defines the operations upon those numbers. In the JVM, floating-point arithmetic is performed on 32-bit floats and 64-bit doubles. For each bytecode that performs arithmetic on floats, there is a corresponding bytecode that performs the same operation on doubles.

A floating-point number has four parts – a sign, a mantissa, a radix, and an exponent. The sign is either a 1 or −1. The mantissa, always a positive number, holds the significant digits of the floating-point number. The exponent indicates the positive or negative power of the radix that the mantissa and sign should be multiplied by. $p = \pm 127$. $1 \le m < 2$.

3.9.1 *Floating-Point Arithmetic*

Floating-point arithmetic derives its name from something that happens when you use exponential notation. Consider the number $A = 123$: it can be written using exponential notation as:

$$A = 1.23 \cdot 10^2 = 12.3 \cdot 10^1 = 123 \cdot 10^0 = 1230 \cdot 10^{-1} \text{ etc.}$$

All of these representations of the number 123 are numerically equivalent. They differ only in their "normalization": where the decimal point appears in the first number. In each case, the number before the multiplication operator ("*") represents the significant figures in the number (which distinguish it from other numbers with the same normalization and exponent); we will call this number the "significand" (also called the "mantissa" in other texts, which call the exponent the "characteristic").

Only two of the representations of the number 123 above are in any kind of standard form. The first representation, $1.23 \cdot 10^2$, is in a form called "scientific notation", and is distinguished by the normalization of the significand: in scientific notation, the significand is always a number greater than or equal to 1 and less than 10.

Standard computer normalization for floating point numbers follows the fourth form in the list above: the significand is greater than or equal to .1, and is always less than 1.

Of course, in a binary computer, all numbers are stored in base 2 instead of base 10; for this reason, the normalization of a binary floating point number simply requires that there be no leading zeroes after the binary point (just as the decimal point separates the 10^0 place from the 10^{-1} place, the binary point separates the 2^0 place from the 2^{-1} place). We will continue to use the decimal number system for our numerical examples, but the impact of the computer's use of the binary number system will be felt as we discuss the way those numbers are stored in the computer. Over the years, floating point formats in computers have not exactly been standardized. While the IEEE (Institute of Electrical and Electronics Engineers) has developed standards in this area, they have not been universally adopted. This is due in large part to the issue of "backwards compatibility": when a hardware manufacturer designs a new computer chip, they usually design it so that programs which ran on their old chips will continue to run in the same way on the new one. Since there was no standardization in floating point formats when the first floating point processing chips (often called "coprocessors" or "FPU"s: "Floating Point Units") were designed, there was no rush among computer designers to conform to the IEEE floating point standards (although the situation has improved with time).

Table 3.4 Floating-point standard for diferent processor

Precision	Sign (# of bits)	Exponent (# of bits)	Significand (# of bits)	Total Length (in bits)	Decimal digits of precision
IEEE/Intel single	1	8	23	32	>6
IEEE single extended	1	≥11	≥32	≥44	>9
IEE/Intel double	1	11	52	64	>15
IEEE double extended	1	≥15	≥64	≥64	>19
Intel internal	1	15	64	80	>19

The following table describes the IEEE standard formats as well as those used in common Intel processors, Table 3.4.

Note first that all of the formats reserve 1 bit to store the sign of the number; this is necessary because the significand is stored as an unsigned fraction in all of these formats (often the first bit of the significand is not even stored, because it is always 1 in a properly normalized floating-point number). The rows describing the IEEE extended formats specify the minimum number of bits which the exponent and significand must have in order to satisfy the standard. The Intel "internal" format is an extended precision format used inside the CPU chip, which allows consecutive floating point operations to be performed with greater precision than that which will eventually be stored.

Examples

We will do all of our examples using decimal, but always keep in mind that the computer always uses binary code.

Example 1.

To find the sum of numbers of $A = 122_{10}$ and $B = 12_{10}$.

We first normalize these numbers as $A = 0.122 \cdot 10^3$ and $B = 0.12 \cdot 10^2$.

We can't simply add two decimal numbers which are multiplied by different exponents. That is, the answers – $0.242 \cdot 10^3$ or $0.242 \cdot 10^2$ are obviously incorrect.

To solve this problem, the number with the smaller exponent must be denormalized before the addition can take place – $B = 0.12 \cdot 10^2$ becomes $B = 0.012 \cdot 10^3$.

Now we can simply add the decimal numbers, since

$$C = A \cdot 10^x + B \cdot 10^x = (A + B) \cdot 10^x,$$

and we get the answer – $C = 0.134 \cdot 10^3$.

Example 2.

$A = 0{,}315290 \cdot 10^{-2}$ and $= 0{,}114082 \cdot 10^{+2}$.

First variant		Second variant	
$A =$	$0{,}315280 \cdot 10^{-2}$	$A = 0{,}000031$	$5280 \cdot 10^{+2}$
$B = 1140$	$0{,}820000 \cdot 10^{-2}$	$B = 0{,}114082$	$\cdot 10^{+2}$
$C =$	$1{,}135280 \cdot 10^{-2}$	$C = 0{,}114114$	$\cdot 10^{+2}$
Incorrect result		**Correct result**	

Example 3.
$A = 0{,}96501 \cdot 10^{+2}$ and $B = 0{,}73004 \cdot 10^{+1}$.

$A =$	$0{,}96501 \cdot 10^{+2}$	
$B =$	$0{,}07300 \cdot 10^{+2}$	
$C =$	$1{,}03801 \cdot 10^{+2}$	← overflow mantissas
$C =$	$1{,}0380 \cdot 10^{+3}$	← **Correct result**

Example 4.
$A = 0{,}24512 \cdot 10^{-8}$ and $B = -0{,}24392 \cdot 10^{-8}$.

$A =$	$0{,}24512 \cdot 10^{-8}$	
$B =$	$-0{,}24392 \cdot 10^{-8}$	
$C =$	$0{,}00120 \cdot 10^{-8} =$	$0{,}12000 \cdot 10^{-10}$

Exponents are commonly stored in these formats as unsigned integers; however, an exponent can be negative as well as positive, and so we must have some technique for representing negative exponents using unsigned integers. This technique is called "biasing": a positive number is added to the exponent before it is stored in to the floating point number. The stored exponent is then called a "biased exponent". If the exponent contains 8 bits, the bias number 127 is added to the exponent before it is stored so that, for example, an exponent of 1 is stored as 128. Since the unsigned exponent can represent numbers between 0 and 255, it should be theoretically possible to store exponents whose values range from −127 to +128 (−127 would stored as the biased exponent value 0, and +128 would be stored as the biased value 255). In practice, the IEEE specification reserves the values 0 and 255, which means that an 8-bit exponent can represent exponent values between −126 and +127. If the stored (biased) exponent has the value 0, and the significand is 0 as well, the value of the floating point number is exactly 0. A floating point number with a stored exponent of 0 and a nonzero significand is of course unnormalized. If the stored exponent has the value 255 (all ones), the floating point number has one of two special meanings:

- if the significand is 0, the number represents infinity, and
- if the significand is not zero, that number represents a "NaN" ("Not a Number"): the result of a division by zero.
- In general, in order to perform any floating point arithmetic operation, the computer must:
- first represent each operand as a normalized number within the limits of its precision (which may result in representation error due to truncation of less significant digits);
- denormalize the smaller of the numbers if an addition or subtraction is being performed (which may again result in representation error due to the denormalization);

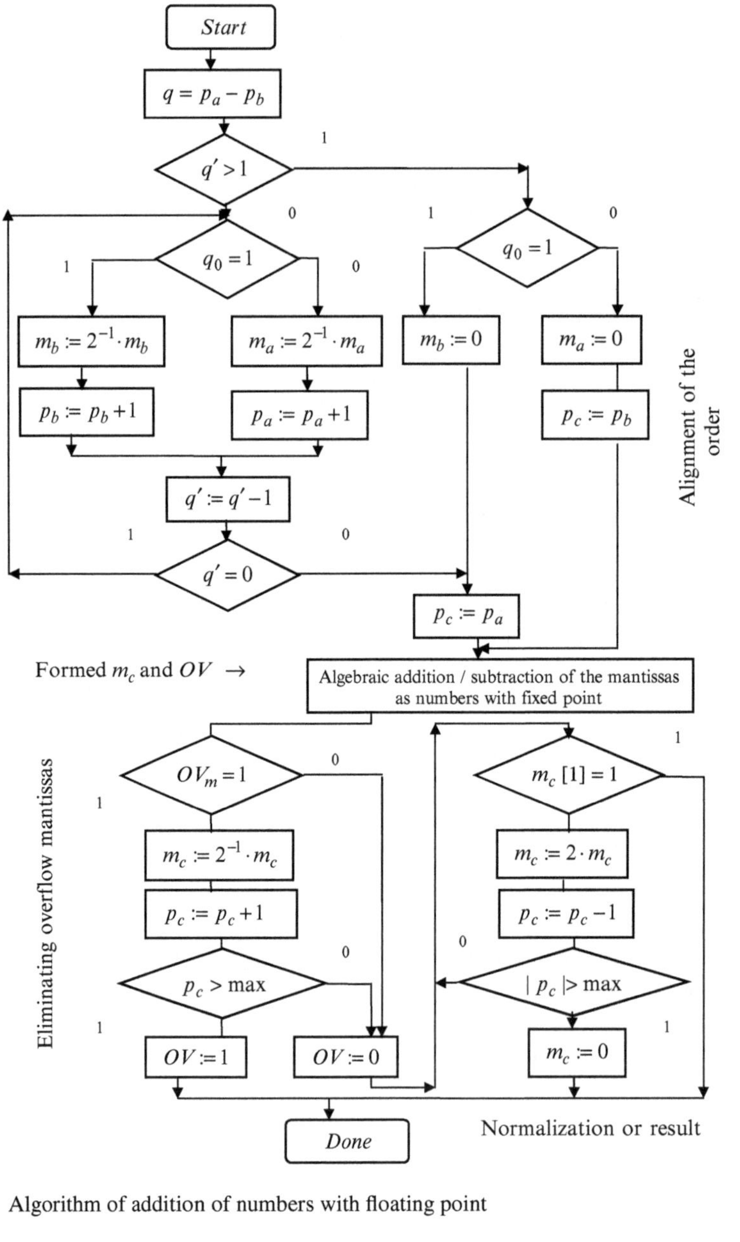

Fig. 3.5 Algorithm of addition of numbers with floating point

- perform the operation (which again may result in representation error due to the finite precision of the floating point processor);
- finally normalize the result.

An analogical algorithm can be used for the operations of multiplication of division for binary numbers, presented in a format with a floating point, Fig. 3.5.

Let

$$A = m_A \cdot q^{p_A};$$

$$B = m_B \cdot q^{p_B};$$

$$C = m_C \cdot q^{p_C};$$

$$D = m_D \cdot q^{p_D}.$$

Then

$$C = A \cdot B = \left(m_A \cdot m_B\right) \cdot q^{p_A + p_B};$$
$$D = A \setminus B = \left(m_A \setminus m_B\right) \cdot q^{p_A - p_B}.$$

Chapter 4
Error Correction in Digital Systems

Abstract This chapter presents error detection and error correction methods that are used in digital systems, such as parity methods, cyclic redundancy check (CRC), Reed-Solomon block and Hamming code. Additionally some of these methods are further explained in few examples enclosed in this chapter.

4.1 Parity Method for Error Detection

The movement of digital data from one location to another can result in transmission errors, the receiver not receiving the same signal as transmitted by the transmitter as a result of electrical noise in the transmission process. Sometimes a noise pulse may be large enough to alter the logic level of the signal. For example, the transmitted sequence 1001 may be incorrectly received as 1101. In order to detect such errors a parity bit is often used. A parity bit is an extra 0 or 1 bit attached to a code group at transmission. In the even parity method the value of the bit is chosen so that the total number of 1s in the code group, including the parity bit, is an even number. For example, in transmitting 1001 the parity bit used would be 0 to give 01001, and thus an even number of 1s. In transmitting 1101 the parity bit used would be 1 to give 11101, and thus an even number of 1s. With odd parity the parity bit is chosen so that the total number of 1s, including the parity bit, is odd. Thus if at the receiver the number of 1s in a code group does not give the required parity, the receiver will know that there is an error and can request that the code group be retransmitted.

4.2 Cyclic Redundancy Check (CRC)

CRC error detection computes the remainder of a polynomial division of a generator polynomial into a message. The remainder, which is usually 16 or 32 bits, is then appended to the message. When another remainder is computed, a nonzero value

B. Borowik et al., *Theory of Digital Automata*, Intelligent Systems, Control and Automation: Science and Engineering 63, DOI 10.1007/978-94-007-5228-3_4,

indicates an error. However, depending on the generator polynomial's size, the process can fail in several ways. It is very difficult to determine how effective a given CRC will be at detecting errors. The probability that a random code word is valid (not detectable as an error), is completely a function of the code rate: $1-2^{-(n-k)}$. Where n is the number of bits of formed from k original bits of data, $(n-k)$ is the number of redundant bits.

Use of the CRC technique for error correction normally requires the ability to send retransmission requests back to the data source.

CRCs are not suitable for protecting against intentional alteration of data. Firstly, as there is no authentication, an attacker can edit a message and recalculate the CRC without the substitution being detected. Secondly, the linear properties of CRC codes allow an attacker even to keep the CRC unchanged while modifying parts of the message.

4.3 Reed-Solomon Block

Reed-Solomon block codes are popular in communications and data-storage applications. Like fire codes, Reed-Solomon-code implementations append symbols to the end of a transmission to locate and correct errors during decoding. Reed-Solomon-code systems' effectiveness at high data rates results from operations taking place at the code-symbol rate or at a fixed number of times per code word. Either way, the number of operations is much smaller than the number of bits. Chips that implement these types of high-speed real-time correctors are commercially available, as are DSP-software options.

Each RS symbol is actually a group of M bits. Just one bit error anywhere in a given symbol spoils the whole symbol. To have fewer bit errors, you'd like to concentrate them into as few RS symbols as possible. That's why RS codes are often called "burst-error-correcting" codes. Many RS codes in use are "shortened" by making the size of the block or the number of used symbols smaller than M (i.e. the size of RS symbol) or smaller than the maximum number of symbols (i.e. 2^{M-1}).

Since RS can be done on any message length and can add any number of extra check symbols, a particular RS code will be expressed as RS $(N,N-R)$ code where N is the total number of symbols per code word; R is the number of check symbols per code word and; therefore, $N-R$ is the number of actual information symbols per code word. The typical RS decoder can correct up to $(N-R)/2$ symbol errors per block.

Generally, a Reed-Solomon corrector provides a number of symbol corrections, say Z, in an N-symbol code word. Z is independent of the location of the errors inside the code word. When you use this method with complex interleaving, this approach can easily correct large error bursts. Further algorithm refinements allow even more correction capability if you know the error location by some other means. These "soft-error" indicators can come, for example, from up-stream decoding violations. It is unable to both locate and correct errors in a block that has more than Z symbol errors.

4.4 Hamming Code

In the late 1940s Claude Shannon was developing information theory and coding as a mathematical model for communication. At the same time, Richard Hamming, a colleague of Shannon's at Bell Laboratories, found a need for error correction in his work on computers. Parity checking was already being used to detect errors in the calculations of the relay-based computers of the day, and Hamming realized that a more sophisticated pattern of parity checking allowed the correction of single errors along with the detection of double errors.

The codes that Hamming devised, the single-error-correcting binary Hamming codes and their single-error-correcting, double-error-detecting extended versions marked the beginning of coding theory. These codes remain important to this day, for theoretical and practical reasons as well as historical.

In a Hamming code, multiple extra bits are computed such that each extra bit will be affected by the data bits in a distinct way. For example, in a system with four data bits, the first extra bit might be affected by a change in data bits 2 thru 4, but not by a change in the first data bit. The second extra bit would be unaffected by an error in the second data bit, and the last extra bit would have no relation to data bit 3. The matrix for this Hamming code would look like Fig. 4.1.

When the extra bits are recalculated at the receiving end, any differences will call out any single corrupted data bit, or indicate an error if two bits are corrupted.

1011 is encoded as 1011 010. If the first bit is corrupted, 0011 010 will be received. Using the first 4 (data) bits to re-calculate the hamming code returns 001. XORing the received extra bits with the calculated extra bits gives us $010 \oplus 001 = 011$ which is the pattern in the extra bits on the top line of the matrix, indicating an error in the first bit. The XOR of the expected error-checking bits with those actually received is called the syndrome of a received code word.

Example 1

When data is transmitted from one location to another there is always the possibility that an error may occur. There are a number of reliable codes that can be used to encode data so that the error can be detected and corrected. With this example you will explore a simple error detection-correction technique called a Hamming Code. A Hamming Code can be used to detect and correct one-bit change in an encoded code word. This approach can be useful as a change in a single bit is more probable than a change in two or more bits.

Consider Fig. 4.2. Data is represented (stored) in every position (1–15) except 1, 2, 4 and 8. These positions (which are powers of 2) are used to store parity (error correction) bits.

d_1	1	0	0	0	0	1	1
d_2	0	1	0	0	1	0	1
d_3	0	0	1	0	1	1	0
d_4	0	0	0	1	1	1	1
	D_1	D_2	D_3	D_4	E_1	E_2	E_3

Fig. 4.1

1	2	3	4	5	6	7	8	9	10	11	12	13	14	15
error correction	error correction	1	error correction	2	3	4	error correction	5	6	7	8	9	10	11

Fig. 4.2

1	2	3	4	5	6	7	8	9	10	11	12	13	14	15
error correction	error correction	**1**	error correction	**0**	**1**	**0**	error correction	**1**	**1**	**0**	**1**	**0**	**1**	**1**

Fig. 4.3

Table 4.1

Positions	$2^0=1$	$2^1=2$	$2^2=4$	$2^3=8$
1	**1**	**0**	**0**	**0**
2	**0**	**1**	**0**	**0**
3	**1**	**1**	**0**	**0**
4	**0**	**0**	**1**	**0**
5	**1**	**0**	**1**	**0**
6	**0**	**1**	**1**	**0**
7	**1**	**1**	**1**	**0**
8	**0**	**0**	**0**	**1**
9	**1**	**0**	**0**	**1**
10	**0**	**1**	**0**	**1**
11	**1**	**1**	**0**	**1**
12	**0**	**0**	**1**	**1**
13	**1**	**0**	**1**	**1**
14	**0**	**1**	**1**	**1**
15	**1**	**1**	**1**	**1**

Using the four parity (error correction bits) positions we can represent 15 values (1–15). These values and their corresponding binary representation are shown in the table below.

Using the format given, data is represented by the 11 non-parity bits. For example the following data item to be encoded (**10101101011**), Fig. 4.3 (Table 4.1):

In positions 3, 6, 9, 10, 12, 14 and 15 we have a '**1**'. Using our previous conversion table we obtain the binary representation for each of these values. We then apply the exclusive OR to the resulting values (essentially setting the parity bit to 1 if an odd # of 1s, else setting it to 0). The results of this activity are shown in Table 4.2:

The parity bits are then put in the proper locations in the table providing the following end result (Fig. 4.4):

This is the encoded code word that would be sent. The receiving side would re-compute the parity bits and compare them to the ones received. If they were the same, no error occurred – if they were different, the location of the flipped bit is determined. For example, let's say that the bit in position 14 was flipped during transmission. The receiving end would see the following encoded sequence (Fig. 4.5):

Table 4.2

	1	1	0	0	3
	0	1	1	0	6
	1	0	0	1	9
	0	1	0	1	10
	0	0	1	1	12
	0	1	1	1	14
	1	1	1	1	15
XOR	1	1	0	1	(11)

1	2	3	4	5	6	7	8	9	10	11	12	13	14	15
1	1	1	0	0	1	0	1	1	1	0	1	0	1	1

Fig. 4.4

1	2	3	4	5	6	7	8	9	10	11	12	13	14	15
1	1	1	0	0	1	0	1	1	1	0	1	0	0	1

Fig. 4.5

Table 4.3

	1	1	0	0	3
	0	1	1	0	6
	1	0	0	1	9
	0	1	0	1	10
	0	0	1	1	12
	1	1	1	1	15
XOR (Fig. 4.5)	1	1	0	0	

Table 4.3 shows the re-calculation at the receiving end.

The re-calculated parity information is then compared to the parity information sent/received. If they are both the same the result (again using an XOR – even parity) will be all 0s. If a single bit is flipped the resulting number will be the position of the errant bit (check back into table).

In the methods of error correction of digital systems we often use logic operation XOR, Figs. 4.6 and 4.7. The following is an example of a program that describes the operating principle of this logical operation.

```
program Summatot_mod2;
{$APPTYPE CONSOLE}
Uses SysUtils;
CONST MRI=10;
VAR A: ARRAY[1..MRI+1] of BYTE;
```

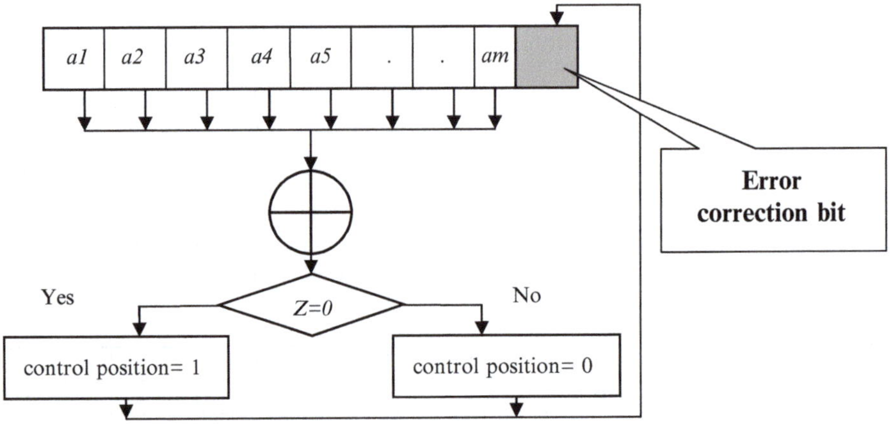

Fig. 4.6 Blok diagram of the error correction

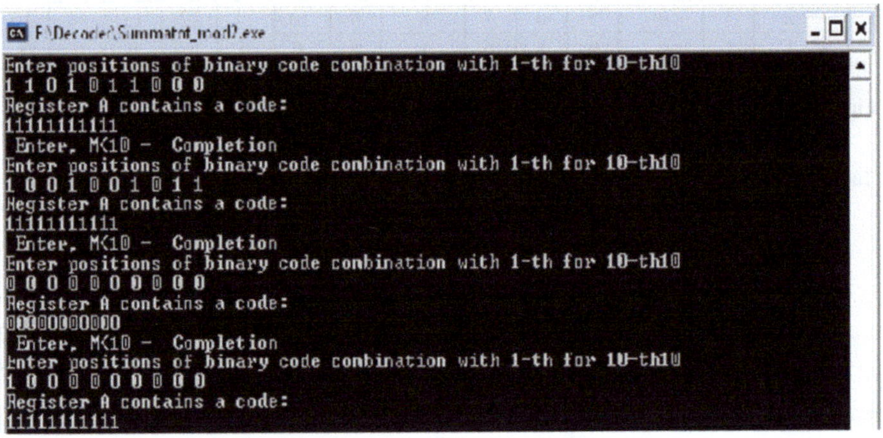

Fig. 4.7 Program that describes the operating principle of error correction

```
I, PZ,X:INTEGER;
begin
REPEAT
WRITELN('Enter positions of binary code combination with 1-th for 10-th',
MRI);
FOR I:=1 TO MRI DO READ(A[I]);
PZ:=0;
FOR I:=1 TO MRI DO PZ:=PZ+A[I];
PZ:=PZ mod 2;
IF ODD(PZ)
THEN A[MRI+1]:=0
ELSE A[MRI+1]:=1;
```

```
WRITELN('Register A contains a code:');
FOR I:=1 TO MRI+1 DO WRITE(A[1]:1, ''); WRITELN;
WRITELN(' Enter, M<10 - Completion');
UNTIL MRI<10
 {TODO -oUser -cConsole Main : Insert code here}
end.
```

In information theory, the Hamming distance between two strings of equal length is the number of positions at which the corresponding symbols are different. Put another way, it measures the minimum number of substitutions required to change one string into the other, or the number of errors that transforms one string into the other ($d_{min} = 3$). For example, 0101 and 0110 has a Hamming distance of two whereas "Butter" and "ladder" are four characters apart.

The Hamming distance is used in digital telecommunication systems, to count the number of flipped bits in a fixed-length binary word as an estimate of error, and therefore is sometimes called the signal distance. Hamming weight analysis of bits is used in several disciplines including information theory, coding theory, and cryptography.

Chapter 5
Boolean Algebra

Abstract The chapter introduces to Boolean Algebra. First the three Boolean operators that are used today are concerned (i.e. AND, OR, NOT). Then the laws of Boolean logic are defined axiomatically using axioms together with theorems, which are presented here in the form of certain equations. The chapter introduces also to the disjunctive normal form (DNF) and to conjunctive normal form (CNF), which allow for the standardization (or normalization) of logical formulas.

Boolean algebra (or Boolean logic) is a logical calculus of truth values, developed by George Boole in the 1840s.

Boolean algebra (symbolic logic) remained dormant until the middle of the Twentieth century. In the 1950s (Huntington) it was used for telephone switching units and the new up and coming electronic computers.

Symbolic logic is used not only in genuinely logical or mathematical domains but also in the natural sciences, and in disciplines such as linguistics, law, and computer technology.

Today Boolean algebra is used every day to help people when doing searches on the Internet. It is more commonly referred to as a Boolean search. The three Boolean operators used today are as follows: AND, OR, NOT.

The laws of Boolean algebra can be defined axiomatically as certain equations called axioms together with their logical consequences called theorems, or semantically as those equations that are true for every possible assignment of 0 or 1 to their variables. The axiomatic approach is sound and complete in the sense that it proves respectively neither more nor fewer laws than the semantic approach.

Boolean algebra is the algebra of two values. These are usually taken to be 0 and 1, as we shall do here, although F and T, false and true, etc. are also in common use. For the purpose of understanding Boolean algebra any Boolean domain of two values will do.

B. Borowik et al., *Theory of Digital Automata*, Intelligent Systems, Control and Automation: Science and Engineering 63, DOI 10.1007/978-94-007-5228-3_5,

5.1 Laws of Boolean Algebra

With values and operations in hand, the next aspect of Boolean algebra is that of laws or properties. As with many kinds of algebra, the principal laws take the form of equations between terms built up from variables using the operations of the algebra. Such an equation is deemed a law or identity only when both sides have the same value for all values of the variables, equivalently when the two terms denote the same operation.

A Boolean algebra can be formally defined as a set **B** of elements *A*, *B*, *C*, *D*,... with the following properties:

1. **B** has two binary operations, ∧ (logical AND) and ∨ (logical OR), which satisfy the idempotent laws

$$A \wedge A = A \quad A \vee A = A, \tag{5.1}$$

the commutative laws

$$A \wedge B = B \wedge A \tag{5.2}$$

$$A \vee B = B \vee A, \tag{5.3}$$

and the associative laws

$$A \wedge (C \wedge B) = C \wedge (B \wedge A) \tag{5.4}$$

$$A \vee (C \vee B) = C \vee (B \vee A). \tag{5.5}$$

2. The operations satisfy the absorption law

$$A \wedge (A \vee B) = A \vee (B \wedge A) = A. \tag{5.6}$$

3. The operations are mutually distributive

$$A \wedge (B \vee C) = (A \wedge B) \vee (A \wedge C) \tag{5.7}$$

$$A \vee (B \wedge C) = (A \vee B) \wedge (A \vee C). \tag{5.8}$$

4. **B** contains universal bounds (the empty set) and (the universal set) which satisfy

$$0 \wedge A = 0 \tag{5.9}$$

$$0 \vee A = A \tag{5.10}$$

$$1 \wedge A = A \tag{5.11}$$

$$1 \vee A = 1. \tag{5.12}$$

5. **B** has a unary operation $A \rightarrow \overline{A}$ of complementation, which obeys the laws

Table 5.1 Specific axioms and theorem of Boolean algebra

(1)	$A=1$, if $A\neq 0$	$A=0$, if $A\neq 1$
(2)	If $A=0$, then $\overline{A}=1$	If $A=1$, then $\overline{A}=0$
(3)	$0+0=0$	$0\cdot 0=0$
(4)	$0+1=1$	$1\cdot 0=0$
(5)	$1+1=1$	$1\cdot 1=1$
(6)	$\overline{0}=1$	$\overline{1}=0$
(7)	$A\vee 0=A$	$A\cdot 1=A;$
(8)	$A\vee 1=1$	$A\cdot 0=0;$
(9)	$A\vee A=A$	$A\cdot A=A;$
(10)	$A\vee\overline{A}=1$	$\overline{A}\cdot A=0$
(11)	$\overline{A+B}=\overline{A}\cdot\overline{B}$	$\overline{A\cdot B}=\overline{A}+\overline{B}$
De Morgan's laws		
(12)	$\overline{A\wedge B}=\overline{A}\vee\overline{B}$	and $\overline{A\vee B}=\overline{A}\wedge\overline{B}$
Double negation		
(13)	$\overline{\left(\overline{A}\right)}=A$	
(14)	$A\cdot B\vee A\cdot\overline{B}=A$	$(A\vee B)(A\vee\overline{B})=A$

$$A\wedge\overline{A}=0 \tag{5.13}$$

$$A\vee\overline{A}=1. \tag{5.14}$$

Huntington (1933) presented the following basis for Boolean algebra:
Commutativity: $A\vee B=B\vee A$.

Associativity: $A\vee(C\vee B)=C\vee(B\vee A)$.

Huntington axiom: $!(!A\vee B)\vee!(!A\vee!B)=A$.

H. Robbins then conjectured that the Huntington axiom could be replaced with the simpler Robbins axiom,

$$!(!(A\vee B)\vee!(A\vee!B))=A.$$

The next table shows that this theory is sufficient to axiomatize all the valid laws or identities of two-valued logic, that is, Boolean algebra. It follows that Boolean algebra as commonly defined in terms of these axioms coincides with the intuitive semantic notion of the valid identities of two-valued logic.

The XOR Gate $A\otimes B=\overline{A}\cdot B+A\cdot\overline{B}$.

Also:

$$A \otimes A = 0;$$
$$A \otimes A \otimes A = A;$$
$$A \otimes \overline{A} = 1;$$
$$A \otimes 1 = \overline{A};$$
$$A \otimes 0 = A;$$
$$A + B = A \oplus B \oplus AB;$$
$$A \cdot B = (A \oplus B) \oplus (A + B).$$

5.2 Disjunctive Normal Form

Soundness follows firstly from the fact that the initial laws or axioms we started from were all identities, that is, semantically true laws. Secondly it depends on the easily verified fact that the rules preserve identities.

Completeness can be proved by first deriving a few additional useful laws and then showing how to use the axioms and rules to prove that a term with n variables, ordered alphabetically say, is equal to its n-ary normal form, namely a unique term associated with the n-ary Boolean operation realized by that term with the variables in that order. It then follows that if two terms denote the same operation (the same thing as being semantically equal), they are both provably equal to the normal form term denoting that operation, and hence by transitivity provably equal to each other. There is more than one suitable choice of normal form, but complete disjunctive normal form will do. A literal is either a variable or a negated variable. A disjunctive normal form (DNF) term is a disjunction of conjunctions of literals. (Associativity allows a term such as $A \vee (B \vee C)$ to be viewed as the ternary disjunction $A \vee B \vee C$, likewise for longer disjunctions, and similarly for conjunction.) A DNF term is complete when every disjunction (conjunction) contains exactly one occurrence of each variable, independently of whether or not the variable is negated. Such a conjunction uniquely represents the operation it denotes by virtue of serving as a coding of those valuations at which the operation returns 1. Each conjunction codes the valuation setting the positively occurring variables to 1 and the negated 1s to 0; the value of the conjunction at that valuation is 1, and hence so is the whole term. At valuations corresponding to omitted conjunctions, all conjunctions present in the term evaluate to 0 and hence so does the whole term.

In Boolean logic, a disjunctive normal form (DNF) is a standardization (or normalization) of a logical formula which is a disjunction of conjunctive clauses. A logical formula is considered to be in DNF if and only if it is a disjunction of one or more conjunctions of one or more literals. A DNF formula is in full disjunctive normal form, if each of its variables appears exactly once in every clause. As in conjunctive normal form (CNF), the only propositional operators in DNF are and,

Table 5.2 CNF and DNF for elementary logic functions

Elementary logic functions	DNF	CNF
0	y	$(x\vee y)\cdot(x\vee\bar{y})\cdot(\bar{x}\vee y)\cdot(\bar{x}\vee\bar{y}$
$x\wedge y$	$x\wedge y$	$(x\vee y)\cdot(x\vee\bar{y})\cdot(\bar{x}\vee y$
$x\wedge\bar{y}$	$x\wedge\bar{y}$	$(x\vee y)\cdot(x\vee\bar{y})\cdot(\bar{x}\vee\bar{y}$
x	$x\cdot\bar{y}\vee x\cdot y$	$(x\vee y)(x\vee\bar{y})$
$\bar{x}\wedge y$	$\bar{x}\wedge y$	$(x\vee y)\cdot(\bar{x}\vee y)\cdot(\bar{x}\vee\bar{y})$
y	$(\bar{x}\cdot y)\vee(x\cdot y)$	$(x\vee y)(\bar{x}\vee y)$
$x\otimes y$	$(\bar{x}\cdot y)\vee(x\cdot\bar{y})$	$(x\vee y)(\bar{x}\vee\bar{y})$
$x\vee y$	$(\bar{x}\cdot y)\vee(x\cdot\bar{y})\vee(x\cdot y)$	$x\vee y$
$x\downarrow y$	$\bar{x}\cdot\bar{y}$	$(x\vee\bar{y})\cdot(\bar{x}\vee y)\cdot(\bar{x}\vee\bar{y}$
$x\equiv y$	$(\bar{x}\cdot\bar{y})\vee(x\cdot y)$	$(x\vee\bar{y})\cdot(\bar{x}\vee y)$
$\bar{y}$	$(\bar{x}\cdot\bar{y})\vee(x\cdot\bar{y})$	$(x\vee\bar{y})\cdot(\bar{x}\vee\bar{y})$
$y\rightarrow x$	$(\bar{x}\cdot\bar{y})\vee(x\cdot\bar{y})\vee(x\cdot y)$	$x\vee\bar{y}$
$\bar{x}$	$(\bar{x}\cdot\bar{y})\vee(\bar{x}\cdot y)$	$(\bar{x}\vee y)\cdot(\bar{x}\vee\bar{y})$
$x\rightarrow y$	$(\bar{x}\cdot\bar{y})\vee(\bar{x}\cdot y)\vee(x\cdot y)$	$\bar{x}\vee y$
$x\mid y$	$(\bar{x}\cdot\bar{y})\vee(x\cdot\bar{y})\vee(\bar{x}\cdot y)$	$\bar{x}\vee\bar{y}$
1	$(x\cdot\bar{y})\vee(\bar{x}\cdot y)\vee(x\cdot y)\vee(\bar{x}\cdot\bar{y}).$	y

or, and not. The not operator can only be used as part of a literal, which means that it can only precede a propositional variable. For example, all of the following formulas are in DNF (Table 5.2):

$$f(A,B,C)=\bar{B}\cdot C\vee A\cdot\bar{B}$$

Or in DNF

$$f(A,B,C)=\bar{B}\cdot C\vee A\cdot\bar{B}=(A\vee\bar{A})\cdot\bar{B}\cdot C\vee A\cdot\bar{B}\cdot(C\vee\bar{C})$$
$$=A\cdot\bar{B}\cdot C\vee\bar{A}\cdot\bar{B}\cdot C\vee A\cdot\bar{B}\cdot\bar{C}.$$

Converting a formula to DNF involves using logical equivalences, such as the double negative elimination, De Morgan's laws, and the distributive law. All logical formulas can be converted into disjunctive normal form. However, in some cases conversion to DNF can lead to an exponential explosion of the formula.

Chapter 6
Basic Logical Functions and Gates. Logic Design

Abstract In this chapter the three fundamental logical operations are reviewed, i.e.. the AND, OR, and NOT. The AND, OR, NOT, NAND, NOR and XOR gates are explained together with their functional implementation. Additionally the chapter presents few combinational digital systems, like a full adder and the seven-segment display and provides examples of designing combinational logic circuits and the way of evaluating logic circuit outputs.

6.1 Basic Logical Functions and Gates

Notation of basic logic gates shown in Tables 6.1 and 6.2.

In some countries such as Russia, Ukraine and others use close to the standard 91–1984 IEEE/ANSI notation of logic elements, see Table 6.2.

While each logical element (Table 6.1) or condition must always have a logic value of either "0" or "1", we also need to have ways to combine different logical signals or conditions to provide a logical result.

These designations we have used in the examples of Chap. 9.

When we deal with logical circuits (as in computers), we not only need to deal with logical functions; we also need some special symbols to denote these functions in a logical diagram. There are three fundamental logical operations, from which all other functions, no matter how complex, can be derived. These functions are named and, or, and not. Each of these has a specific symbol and a clearly-defined behavior, as follows:

6.1.1 *The NOT Gate, or Inverter*

The inverter is a little different from AND and OR gates in that it always has exactly one input as well as one output. Whatever logical state is applied to the input, the opposite state will appear at the output.

B. Borowik et al., *Theory of Digital Automata*, Intelligent Systems, Control and Automation: Science and Engineering 63, DOI 10.1007/978-94-007-5228-3_6,

Table 6.1

91-1984 IEEE/ANSI	**The traditional notation of logic elements**	**Logical function**
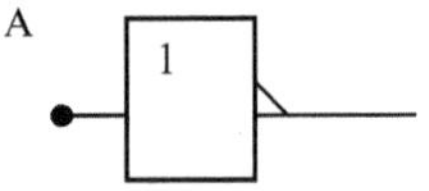	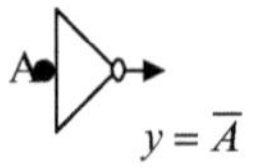	The NOT gate, or inverter $y = \overline{A}$
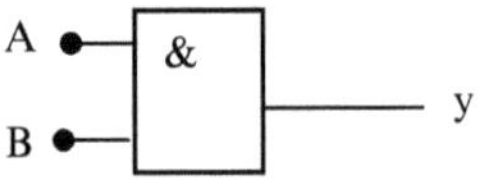	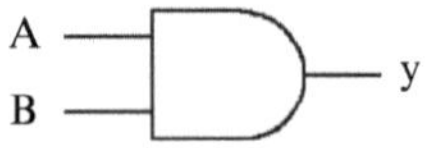	The AND gate $y = A \cdot B = A \wedge B = A \& B$
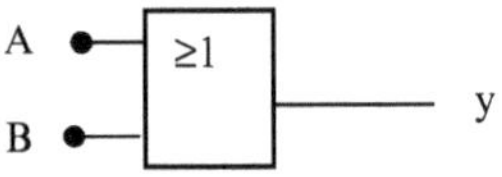	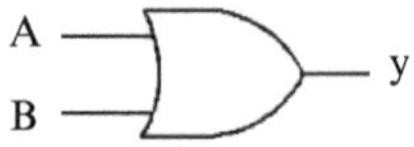	The OR gate $y = A + B = A \vee B$
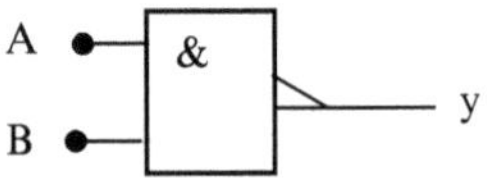	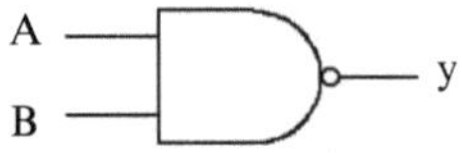	The NAND gate $y = \overline{A \cdot B} = \overline{A} + \overline{B}$
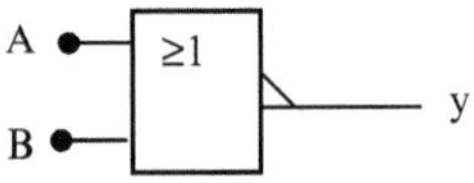	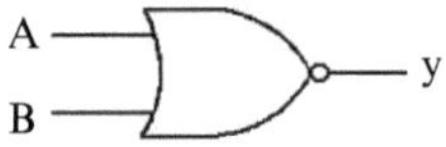	The NOR gate $y = \overline{A + B} = \overline{A} \cdot \overline{B}$
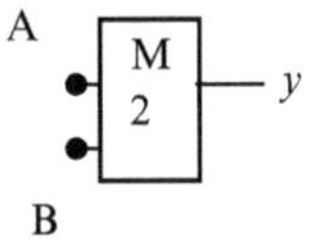	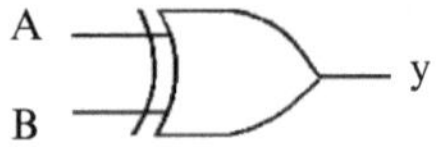	The XOR gate $y = A \otimes B = \overline{A}B + A\overline{B}$

The NOT function is denoted by a horizontal bar over the value to be inverted, as shown in the figure in Table 6.1. In some cases a single quote mark (') may also be used for this purpose: $A = 0$ and $A' = 1$ $(0' = 1)$ and $1' = 0$. For greater clarity in some logical expressions, we will use the overbar most of the time.

In the inverter symbol, the triangle actually denotes only an amplifier, which in digital terms means that it "cleans up" the signal but does not change its logical sense. It is the circle at the output which denotes the logical inversion. The circle could have been placed at the input instead, and the logical meaning would still be the same.

6.1.2 The AND Gate

The AND gate implements the AND function. Both inputs must have logic 1 signals applied to them in order for the output to be a logic 1. With either input at logic 0, the output will be held to logic 0.

There is no limit to the number of inputs that may be applied to an AND function, so there is no functional limit to the number of inputs an AND gate may have.

Table 6.2

Basic logical gates	**Notation**	**Examples of the operation of gates (for AND, NOT, OR)**
The AND gate	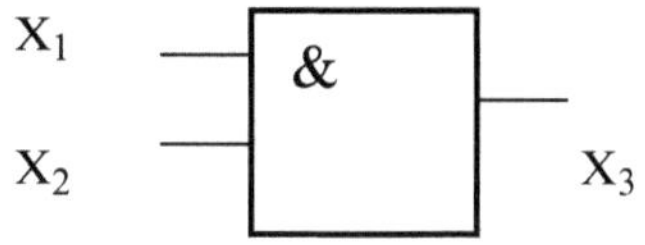	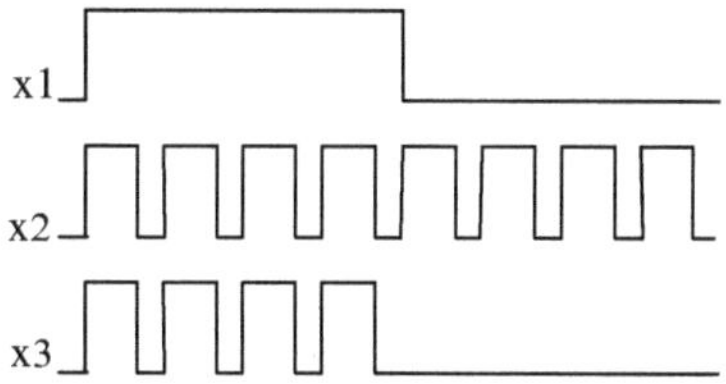
The NOT gate, or inverter	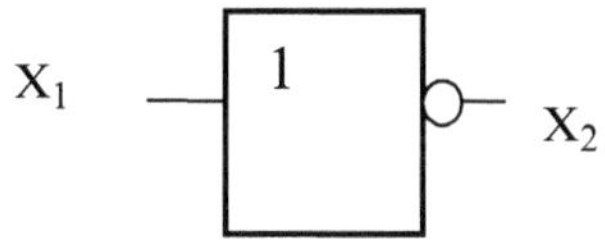	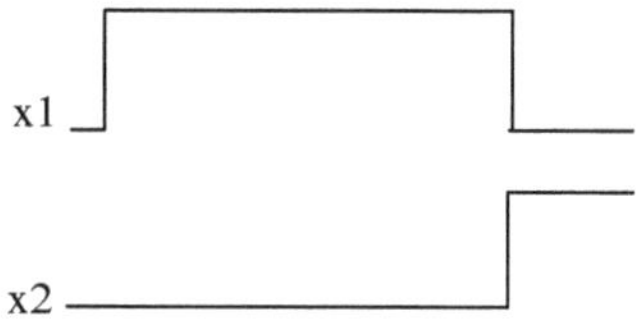
The OR gate	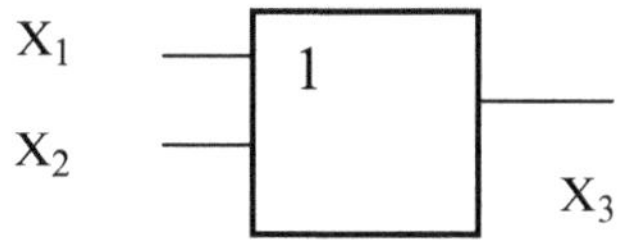	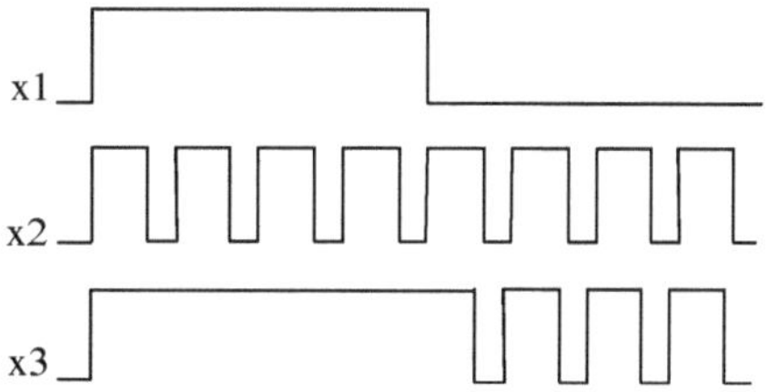
The NOR gate	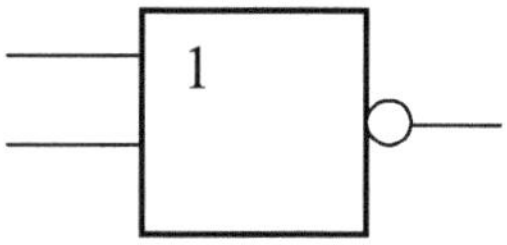	
The NAND gate		
The XOR gate	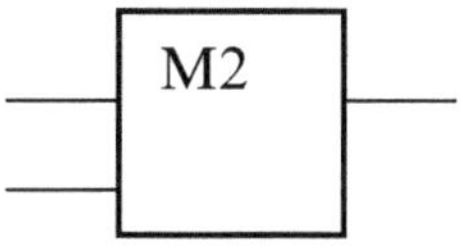	

However, for practical reasons, commercial AND gates are most commonly manufactured with 2, 3, or 4 inputs.

ICs were first developed in the 1960s. They are densely populated miniature electronic circuits made up of hundreds and sometimes thousands of microscopically small transistors, resistors, diodes and capacitors, all connected together on a single chip of silicon.

When assembled in a single package, as shown in Fig. 6.1, we call the device an IC.

There are two broad groups of IC: digital ICs and linear ICs. Digital ICs contain simple switching-type circuits used for logic control and calculators, linear ICs incorporate amplifier-type circuits which can respond to audio and radio frequency

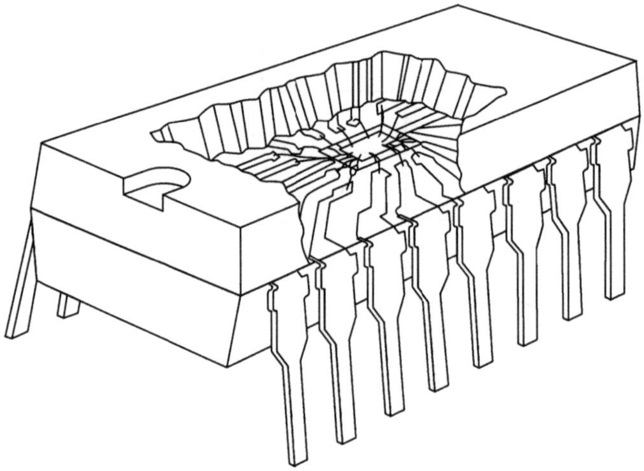

Fig. 6.1 IC

signals. The most versatile linear IC is the operational amplifier which has applications in electronics, instrumentation and control.

The IC is an electronic revolution. ICs are more reliable, cheaper and smaller than the same circuit made from discrete or separate transistors, and electronically superior. One IC behaves differently than another because of the arrangement of the transistors within the IC.

Manufacturers' data sheets describe the characteristics of the different ICs, which have a reference number stamped on the top.

When building circuits, it is necessary to be able to identify the IC pin connection by number. The number 1 pin of any IC is indicated by a dot pressed into the encapsulation; it is also the pin to the left of the cutout. Since the packaging of ICs has two rows of pins they are called DIL (dual in line) packaged ICs.

ICs are sometimes connected into DIL sockets and at other times are soldered directly into the circuit.

A standard Integrated Circuit (IC) package contains 14 or 16 pins, for practical size and handling. A standard 14-pin package can contain four 2-input gates, three 3-input gates, or two 4-input gates, and still have room for two pins for power supply connections, Figs. 6.2 and 6.3.

6.1.3 The OR Gate

The OR gate is sort of the reverse of the AND gate. The OR function, like its verbal counterpart, allows the output to be true (logic 1) if any one or more of its inputs are true. In symbols, the OR function is designated with a plus sign (+). In logical diagrams, the symbol to the left designates the OR gate.

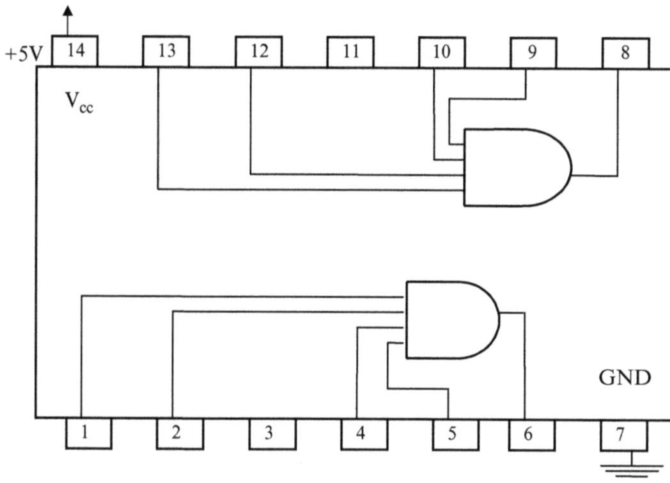

Fig. 6.2 IC 7421

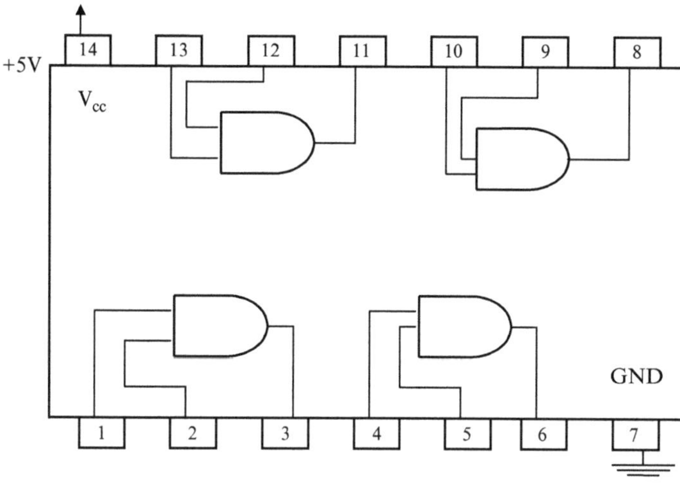

Fig. 6.3 IC 74LS08

As with the AND function, the OR function can have any number of inputs. However, practical commercial OR gates are mostly limited to 2, 3, and 4 inputs, as with AND gates.

While the three basic functions AND, OR, and NOT are sufficient to accomplish all possible logical functions and operations, some combinations are used so commonly that they have been given names and logic symbols of their own.

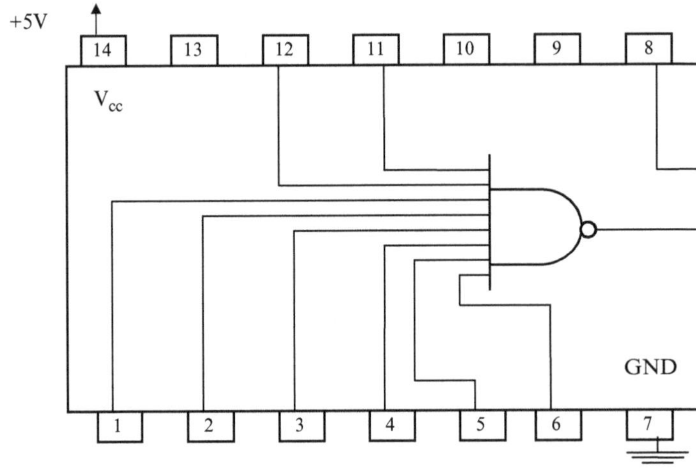

Fig. 6.4 IC 7430

We will discuss three of these. The first is called NAND, and consists of an AND function followed by a NOT function. The second, as you might expect, is called NOR. This is an OR function followed by NOT. The third is a variation of the OR function, called the Exclusive-OR, or XOR function. As with the three basic logic functions, each of these derived functions has a specific logic symbol and behavior, which we can summarize as follows:

6.1.4 The NAND Gate

The NAND gate implements the NAND function, which is exactly inverted from the AND function you already examined. With the gate shown to the left, both inputs must have logic 1 signals applied to them in order for the output to be a logic 0. With either input at logic 0, the output will be held to logic 1.

The circle at the output of the NAND gate denotes the logical inversion, just as it did at the output of the inverter. Also in the figure (Table 6.1), note that the overbar is a solid bar over both input values at once. This shows that it is the AND function itself that is inverted, rather than each separate input.

As with AND, there is no limit to the number of inputs that may be applied to a NAND function, so there is no functional limit to the number of inputs a NAND gate may have. However, for practical reasons, commercial NAND gates are most commonly manufactured with 2, 3, 4 or 8 inputs, to fit in a 14-pin or 16-pin package, Figs. 6.4 and 6.5.

Fig. 6.5 IC 7400

6.1.5 *The NOR Gate*

The NOR gate is an OR gate with the output inverted. Where the OR gate allows the output to be true (logic 1) if any one or more of its inputs are true, the NOR gate inverts this and forces the output to logic 0 when any input is true.

In symbols, the NOR function is designated with a plus sign (+), with an overbar over the entire expression to indicate the inversion. In logical diagrams, the symbol to the left designates the NOR gate. As expected, this is an OR gate with a circle to designate the inversion.

The NOR function can have any number of inputs, but practical commercial NOR gates are mostly limited to 2, 3, and 4 inputs, as with other gates in this class, to fit in standard IC packages.

6.1.6 *The Exclusive-OR, or XOR Gate*

The Exclusive-OR, or XOR function is an interesting and useful variation on the basic OR function. Verbally, it can be stated as, "Either A or B, but not both ($y = A \otimes B = \overline{A}B + A\overline{B}$)." The XOR gate produces a logic 1 output only if its two inputs are different. If the inputs are the same, the output is a logic 0.

The XOR symbol is a variation on the standard OR symbol. It consists of a plus (+) sign with a circle around it. The logic symbol, as shown here, is a variation on the standard OR symbol.

Unlike standard OR/NOR and AND/NAND functions, the XOR function always has exactly two inputs, and commercially manufactured XOR gates are the same. Four XOR gates fit in a standard 14-pin IC package.

6.2 Universal Gates

Universal gates are the ones which can be used for implementing any gate like AND, OR and NOT, or any combination of these basic gates; NAND and NOR gates are universal gates. But there are some rules that need to be followed when implementing NAND or NOR based gates.

To facilitate the conversion to NAND and NOR logic, we have two new graphic symbols for these gates (Appendix D).

NAND Gate (Fig. 6.6)

NOR Gate (Fig. 6.7)

Any logic function can be implemented using NAND gates. To achieve this, first the logic function has to be written in Sum of Product (SOP) form. Once logic function is converted to SOP, then is very easy to implement using NAND gate. In other words any logic circuit with AND gates in first level and OR gates in second level can be converted into a NAND-NAND gate circuit (Fig. 6.8).

Implementing AND using NAND gates (Fig. 6.9)

Implementing OR using NAND gates (Fig. 6.10)

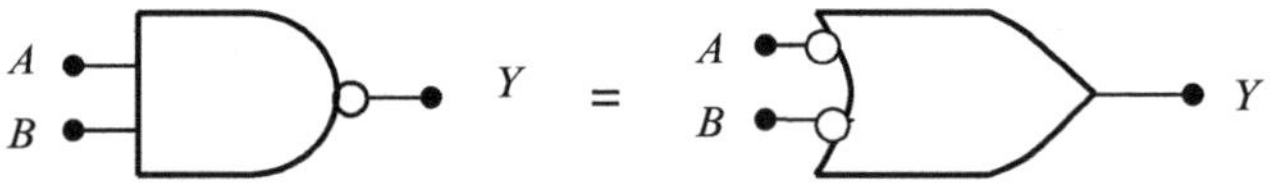

$Y = \overline{A \cdot B} = \overline{A} + \overline{B}$ $\qquad$ $Y = \overline{A} + \overline{B} = \overline{A \cdot B}$

Fig. 6.6

NOR Gate

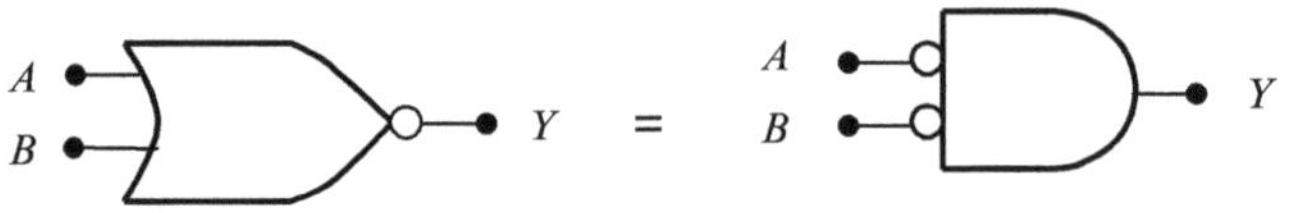

$Y = \overline{A + B} = \overline{A} \cdot \overline{B}$ $\qquad$ $Y = \overline{A} \cdot \overline{B} = \overline{A + B}$

Fig. 6.7

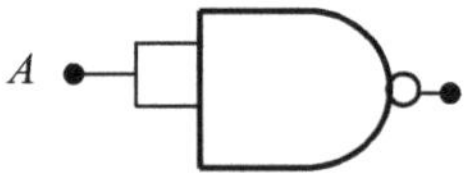

$$Y = \overline{A \cdot A} = \overline{A}$$

Fig. 6.8

Implementing AND using NAND gates

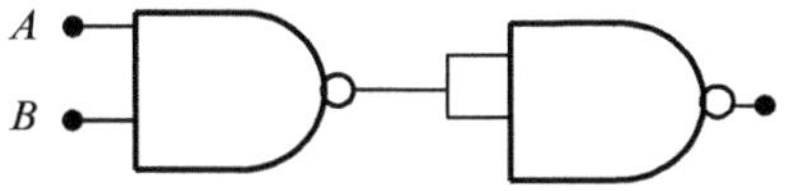

$$Y=\overline{\overline{(A\cdot B)}\cdot\overline{(A\cdot B)}}=A\cdot B$$

Fig. 6.9

Implementing OR using NAND gates

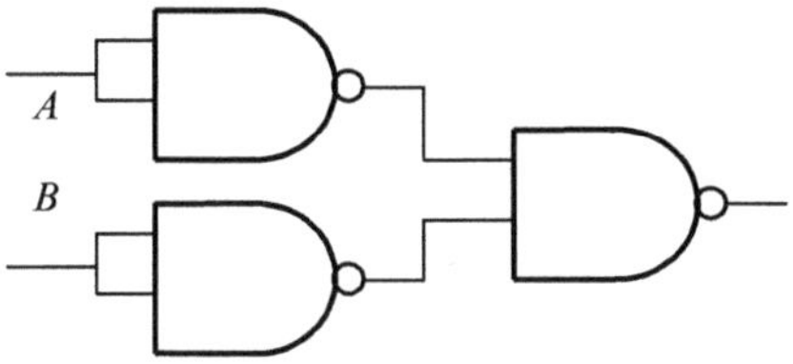

$$Y=\overline{\overline{(A\cdot A)}\cdot\overline{(B\cdot B)}}=$$

$$=\overline{\bar{A}\cdot\bar{B}}=A+B$$

Fig. 6.10

Implementing an inverter using NOR gate

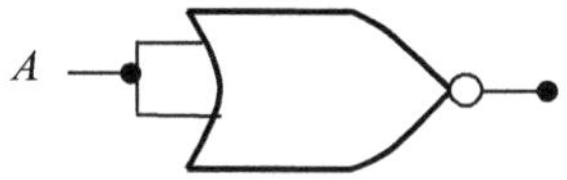

$$Y=\overline{A+A}=\bar{A}$$

Fig. 6.11

6.2.1 Realization of Logic Function Using NOR Gates

Any logic function can be implemented using NAND gates. To achieve this, first the logic function has to be written in Sum of Product (SOP) form. Once logic function is converted to SOP, then is very easy to implement using NAND gate. In other words any logic circuit with AND gates in first level and OR gates in second level can be converted into a NAND-NAND gate circuit.

Implementing an inverter using NOR gate (Fig. 6.11)

Implementing AND using NOR gates (Fig. 6.12)

Implementing OR using NOR gates (Fig. 6.13)

The next section of this book is devoted to combinational logic and deals with various aspects of the analysis and design of combinational switching circuits. The particular characteristic of a combinational switching circuit is that its outputs are functions of only the present circuit inputs. First, switching algebra is introduced as the basic mathematical tool essential for dealing with problems encountered in the study of switching circuits. Switching expressions are defined and are found to be instrumental in describing the logical properties of switching circuits. Systematic

Implementing AND using NOR gates

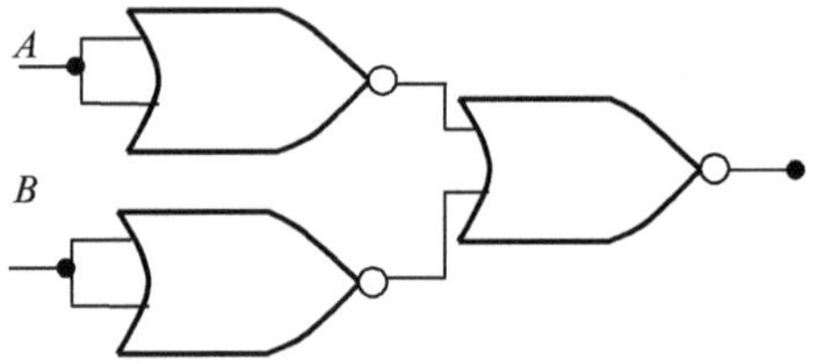

$$Y = \overline{(\overline{A+A}) + (\overline{B+B})} =$$
$$= \overline{\overline{A} + \overline{B}} = A \cdot B$$

Fig. 6.12

Implementing OR using NOR gates

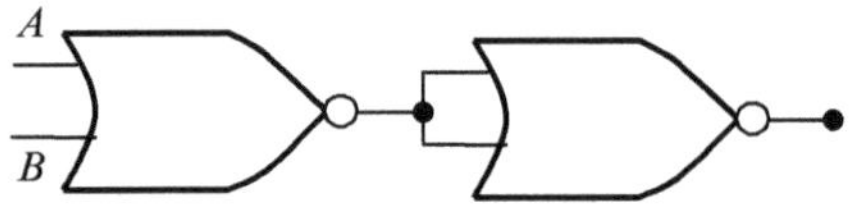

$$Y = \overline{(\overline{A+B}) + (\overline{A+B})} =$$
$$= A + B$$

Fig. 6.13

simplification procedures of these expressions are next presented; these lead to more economical circuits. Logical design is studied with special attention to conventional logic, complementary metaloxide semiconductor (CMOS) circuits, and threshold logic.

6.3 Combinational Logic Circuits

To choose representations, engineers consider types of digital systems. Most digital systems divide into "combinational systems" and "sequential systems." A combinational system always presents the same output when given the same inputs. It is basically a representation of a set of logic functions, as already discussed.

A sequential system is a combinational system with some of the outputs fed back as inputs. This makes the digital machine perform a "sequence" of operations. Unlike Sequential Logic Circuits whose outputs are dependant on both their present inputs and their previous output state giving them some form of Memory, the outputs of Combinational Logic Circuits are only determined by the logical function of their current input state, logic "0" or logic "1", at any given instant in time as they have no feedback, and any changes to the signals being applied to their inputs will immediately have an effect at the output. In other words, in a Combinational Logic Circuit, the output is dependant at all times on the combination of its inputs and if one of its inputs condition changes state so does the output as combinational circuits have "no memory", "timing" or "feedback loops" (Fig. 6.14).

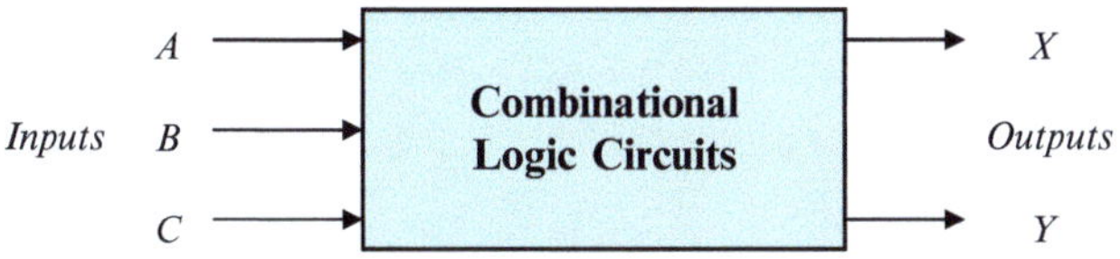

Fig. 6.14

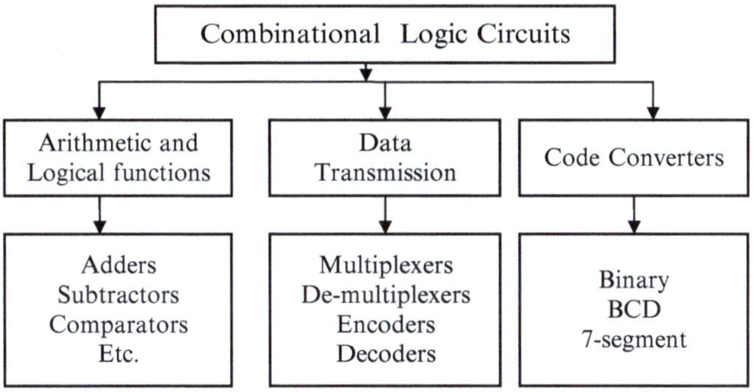

Fig. 6.15

Combination Logic Circuits are made up from basic logic NAND, NOR or NOT gates that are "combined" or connected together to produce more complicated switching circuits. These logic gates are the building blocks of combinational logic circuits. An example of a combinational circuit is a decoder, which converts the binary code data present at its input into a number of different output lines, one at a time producing an equivalent decimal code at its output.

Combinational logic circuits can be very simple or very complicated and any combinational circuit can be implemented with only NAND and NOR gates as these are classed as "universal" gates. The three main ways of specifying the function of a combinational logic circuit are:

Truth Table: provides a concise list that shows the output values in tabular form for each possible combination of input variables.
Boolean Algebra: forms an output expression for each input variable that represents a logic "1".
Logic Diagram: shows the wiring and connections of each individual logic gate that implements the circuit.

As combination logic circuits are made up from individual logic gates only, they can also be considered as "decision making circuits" and combinational logic is about combining logic gates together to process two or more signals in order to produce at least one output signal according to the logical function of each logic gate. Common combinational circuits made up from individual logic gates that carry out a desired application include Multiplexers, De-multiplexers, Encoders, Decoders, Full and Half Adders etc., Fig. 6.15.

Fig. 6.16

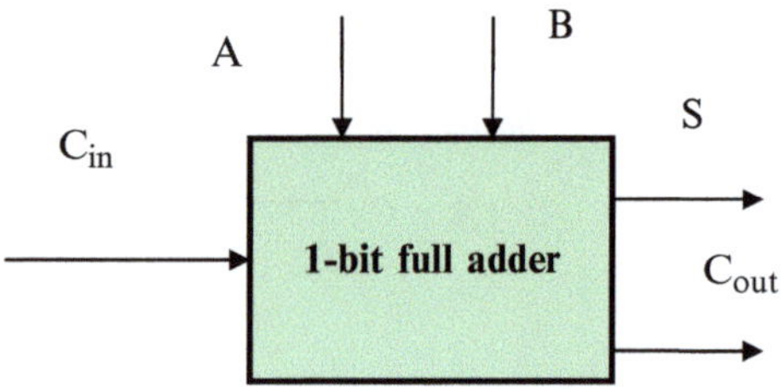

Table 6.3

Inputs			Outputs	
A	B	C_{in}	S	C_{out}
0	0	0	0	0
0	0	1	1	0
0	1	0	1	0
0	1	1	0	1
1	0	0	1	0
1	0	1	0	1
1	1	0	0	1
1	1	1	1	1

The principal application of switching theory is in the design of digital circuits. The design of such circuits is commonly referred to as logical (or logic) design. Most digital systems are constructed from electronic switching circuits. In this section, we describe some components that are typical of the basic building blocks used in constructing digital systems. Switching algebra will be used to describe the logical behavior of networks composed of these building blocks as well as to manipulate and simplify switching expressions, thereby reducing the number of components used in the design. We shall be concerned with the logic functions that a circuit performs rather than with its electronic structure or behavior. These examples will introduce us to some practical aspects of logic design in which the speed of operation and area limitations require ingenuity in arriving at a proper compromise.

6.4 Full Adder

A full adder adds binary numbers and accounts for values carried in as well as out. A one-bit full adder adds three one-bit numbers, often written as *A*, *B*, and C_{in}; *A* and *B* are the operands, and C_{in} is a bit carried in, Fig. 6.16.

The circuit produces a two-bit output sum typically represented by the signals C_{out} and *S* (Table 6.3).

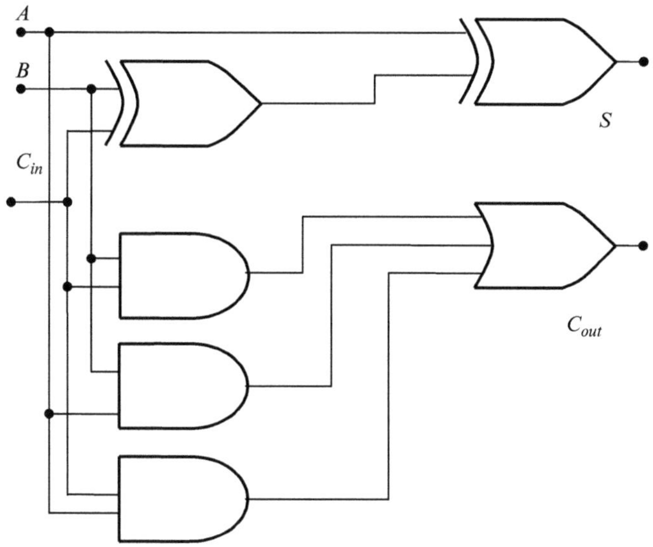

Fig. 6.17

This truth table translates to the logical relationship

Output S

$$S = \overline{A} \bullet \overline{B} \bullet C_{in} + \overline{A} \bullet B \bullet \overline{C}_{in} + A \bullet \overline{B} \bullet \overline{C}_{in} + A \bullet B \bullet C_{in} =$$
$$= \overline{A} \bullet \left(\overline{B} \bullet C_{in} + B \bullet \overline{C}_{in}\right) + A\left(\overline{B} \bullet \overline{C}_{in} + B \bullet C_{in}\right) = \overline{A} \bullet \left(B \otimes C_{in}\right) + A \bullet \left(\overline{B \otimes C_{in}}\right) =$$
$$= A \otimes \left[B \otimes C_{in}\right].$$

The carry bit output is given by the relationship

Output C_{out}

$$C_{out} = \overline{A} \cdot B \cdot C_{in} + A \cdot \overline{B} \cdot C_{in} + A \cdot B \cdot \overline{C}_{in} + A \cdot B \cdot C_{in} =$$
$$= B \cdot C_{in} \cdot (\overline{A} + A) + A \cdot C_{in} \cdot (\overline{B} + B) + A \cdot B \cdot (\overline{C}_{in} + C_{in}) =$$
$$= B \cdot C_{in} + A \cdot C_{in} + A \cdot B.$$

Thus by design, we have the following logic circuit, Fig. 6.17. Figure 6.18 shows an example of the adder circuit in Multisim (Adder circuit in MC8 – Appendix B).

6.5 Seven-Segment Display

A popular method for displaying decimal digits is by means of the seven-segment display shown in Fig. 6.19. The display consists of a BCD-to-seven-segment decoder and seven separate light segments (usually light-emitting diodes or crystals) each

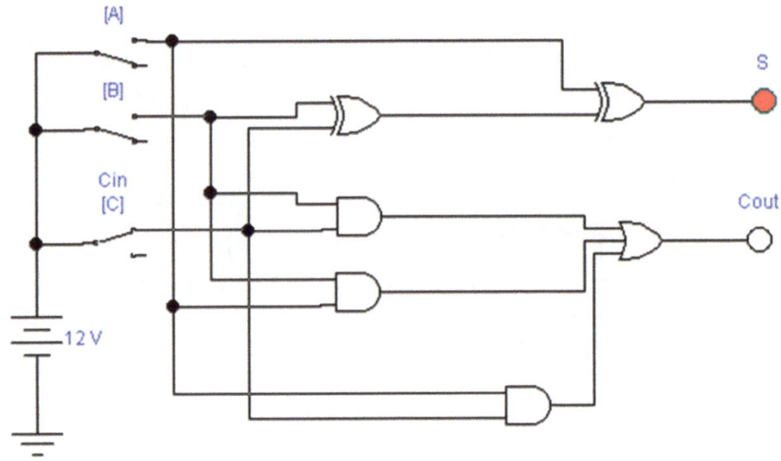

Fig. 6.18

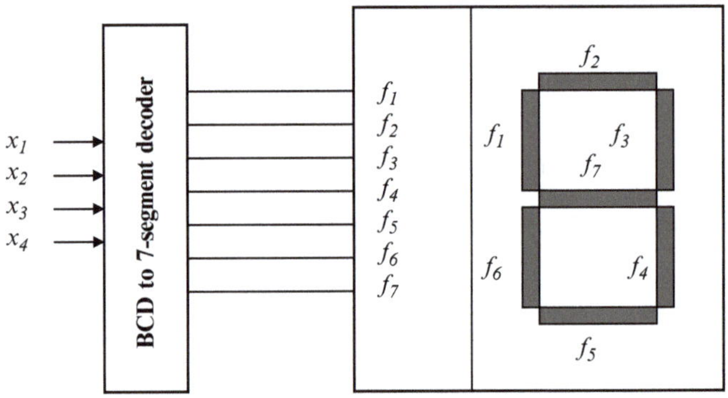

Fig. 6.19

of which can be turned on and off independently of the others. The display receives its inputs in the form of BCD coded digits and transforms these inputs to obtain the pattern of the corresponding decimal digit. Table 6.4 can be viewed as the truth table for the output functions of the BCD-to-seven-segment decoder. The seven-segment code corresponding to each digit is directly obtained from the pattern. For example, to display the decimal digit 2, segments f_2, f_3, f_5, f_6, f_7 are turned on while segments f_4 and f_1 remain off. In a similar manner, the rest of the seven-segment code is obtained. The segment excitation functions can now be determined directly from the table or by using maps. The expressions for the segment excitation functions are thus as follows:

Table 6.4

Decimal digit	x_1	x_2	x_3	x_4	f_1	f_2	f_3	f_4	f_5	f_6	f_7
1	0	0	0	1	0	0	1	1	0	0	0
2	0	0	1	0	0	1	1	0	1	1	1
3	0	0	1	1	0	1	1	1	1	0	1
4	0	1	0	0	1	0	1	1	0	0	1
5	0	1	0	1	1	1	0	1	1	0	1
6	0	1	1	0	1	0	0	1	1	1	1
7	0	1	1	1	0	1	1	1	0	0	0
8	1	0	0	0	1	1	1	1	1	1	1
9	1	0	0	1	1	1	1	1	0	0	1
0	0	0	0	0	1	1	1	1	1	1	0

$$f_1 = x_1 \vee x_2\cdot\bar{x}_3 \vee x_2\cdot\bar{x}_4 \vee \bar{x}_3\cdot\bar{x}_4.$$

$$f_2 = x_1 \vee x_3\cdot x_4 \vee x_2\cdot x_4 \vee \bar{x}_2\cdot\bar{x}_4.$$

$$f_3 = \bar{x}_2 \vee x_3\cdot x_4 \vee \bar{x}_3\cdot\bar{x}_4.$$

$$f_4 = \bar{x}_3 \vee x_2 \vee x_4.$$

$$f_5 = \bar{x}_2\cdot\bar{x}_4 \vee \bar{x}_2\cdot x_3 \vee x_3\cdot\bar{x}_4 \vee x_2\cdot\bar{x}_3\cdot x_4.$$

$$f_6 = \bar{x}_2\cdot\bar{x}_4 \vee x_3\cdot\bar{x}_4.$$

$$f_7 = x_1 \vee \bar{x}_2\cdot x_3 \vee x_2\cdot\bar{x}_3 \vee x_3\cdot\bar{x}_4.$$

For f_1 logic circuit, Fig. 6.20 (logic gates: NOT, AND, OR)
For f_1 logic circuit, Fig. 6.21 (logic gate: NAND)
Checking of the final logic circuit device, Fig. 6.22.

6.6 Design Combinational Logic Circuits

Example 1.

Necessary to design a logic circuit which receives input signals *A,B,C,D* from analog-to-digital converter (ADC), Fig. 6.23.

The resolution of the converter indicates the number of discrete values it can produce over the range of analog values. The values are usually stored electronically in binary

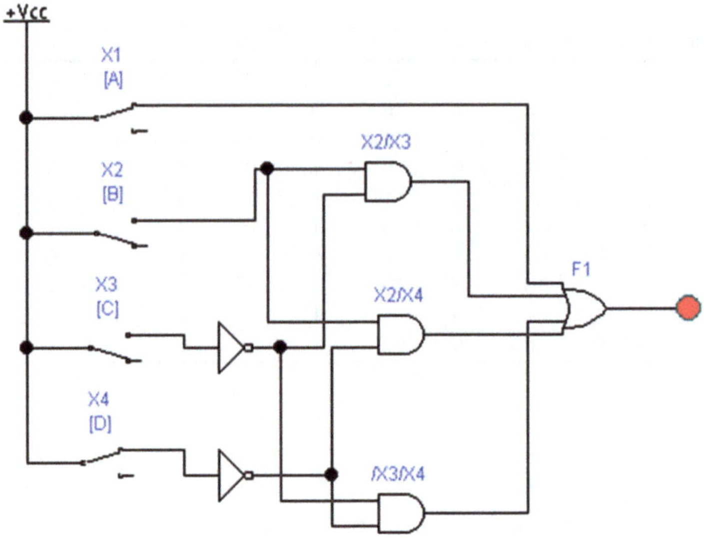

Fig. 6.20

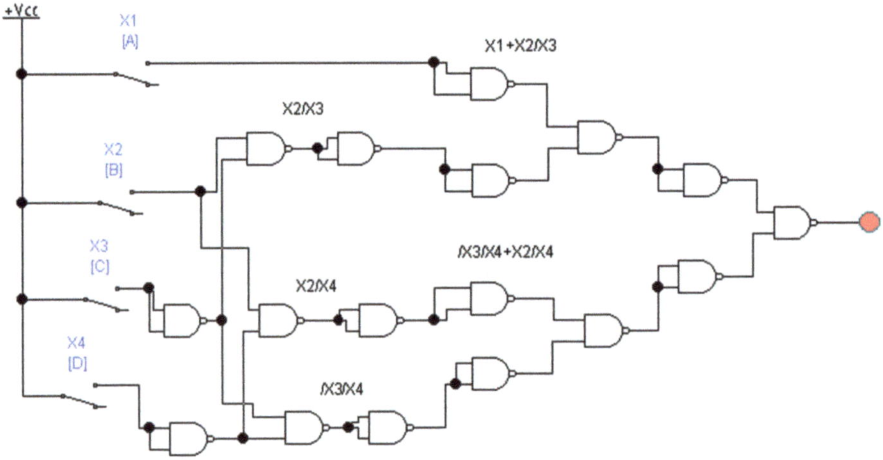

Fig. 6.21

form, so the resolution is usually expressed in bits. In consequence, the number of discrete values available, or "levels", is usually a power of 2. Logical function Z has a high level ($Z=1$) for the following values ABCD: 0111_2; 1000_2; 1001_2; 1010_2; 1011_2; 1100_2; 1101_2; 1110_2; 1111_2.

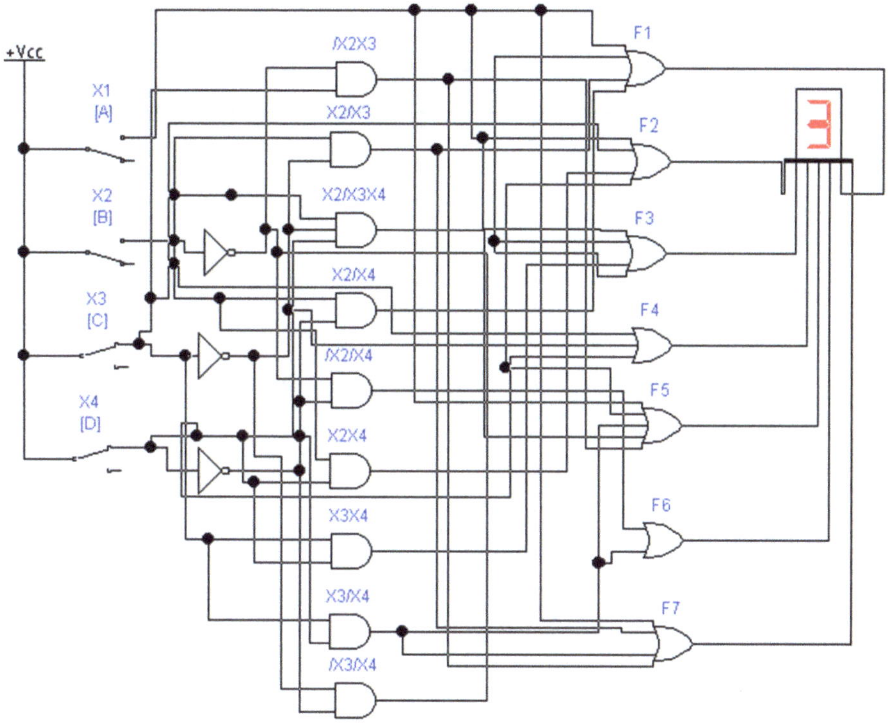

Fig. 6.22

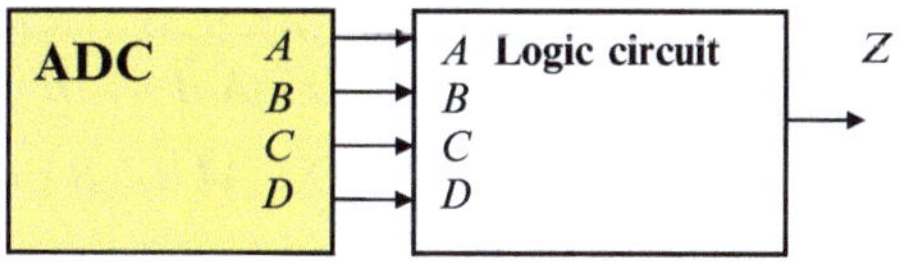

Fig. 6.23

Step 1. We write the truth Table 6.5.

Step 2. From the truth table choose a value of logic function Z, at which it takes the value 1.

$$Z = \overline{A}{\cdot}B{\cdot}C{\cdot}D + A{\cdot}\overline{B}{\cdot}\overline{C}{\cdot}\overline{D} + A{\cdot}\overline{B}{\cdot}\overline{C}{\cdot}D + A{\cdot}\overline{B}{\cdot}C{\cdot}\overline{D} + A{\cdot}\overline{B}{\cdot}C{\cdot}D + \\ A{\cdot}B{\cdot}\overline{C}{\cdot}\overline{D} + A{\cdot}B{\cdot}\overline{C}{\cdot}D + A{\cdot}B{\cdot}C{\cdot}\overline{D} + A{\cdot}B{\cdot}C{\cdot}D.$$

Table 6.5

№	Inputs				Outputs	Minterms
	A	B	C	D	Z	
0	0	0	0	0	0	
1	0	0	0	1	0	
2	0	0	1	0	0	
3	0	0	1	1	0	
4	0	1	0	0	0	
5	0	1	0	1	0	
6	0	1	1	0	0	
7	0	1	1	1	1	$\rightarrow \overline{A}{\cdot}B{\cdot}C{\cdot}D$
8	1	0	0	0	1	$\rightarrow A{\cdot}\overline{B}{\cdot}\overline{C}{\cdot}\overline{D}$
9	1	0	0	1	1	$\rightarrow A{\cdot}\overline{B}{\cdot}\overline{C}{\cdot}D$
10	1	0	1	0	1	$\rightarrow A{\cdot}\overline{B}{\cdot}C{\cdot}\overline{D}$
11	1	0	1	1	1	$\rightarrow A{\cdot}\overline{B}{\cdot}C{\cdot}D$
12	1	1	0	0	1	$\rightarrow A{\cdot}B{\cdot}\overline{C}{\cdot}\overline{D}$
13	1	1	0	1	1	$\rightarrow A{\cdot}B{\cdot}\overline{C}{\cdot}D$
14	1	1	1	0	1	$\rightarrow A{\cdot}B{\cdot}C{\cdot}\overline{D}$
15	1	1	1	1	1	$\rightarrow A{\cdot}B{\cdot}C{\cdot}D$

Step 3. Minimization of the logic function

$$
\begin{aligned}
z &= \overline{A}{\cdot}B{\cdot}C{\cdot}D + A{\cdot}\overline{B}{\cdot}\overline{C}{\cdot}\left(\overline{D}+D\right) + A{\cdot}\overline{B}{\cdot}C{\cdot}\left(\overline{D}+D\right) + \\
&+ A{\cdot}B{\cdot}\overline{C}{\cdot}\left(\overline{D}+D\right) + A{\cdot}B{\cdot}C{\cdot}\left(\overline{D}+D\right) = \\
&= \overline{A}{\cdot}B{\cdot}C{\cdot}D + A{\cdot}\overline{B}{\cdot}\overline{C} + A{\cdot}\overline{B}{\cdot}C + A{\cdot}B{\cdot}\overline{C} + A{\cdot}B{\cdot}C = \\
&= \overline{A}{\cdot}B{\cdot}C{\cdot}D + A{\cdot}\overline{B}{\cdot}\left(\overline{C}+C\right) + A{\cdot}B{\cdot}\left(\overline{C}+C\right) = \\
&= \overline{A}{\cdot}B{\cdot}C{\cdot}D + A{\cdot}\overline{B} + A{\cdot}B = \\
&= \overline{A}{\cdot}B{\cdot}C{\cdot}D + A{\cdot}\left(\overline{B}+B\right) = \\
&= \overline{A}{\cdot}B{\cdot}C{\cdot}D + A = B{\cdot}C{\cdot}D + A.
\end{aligned}
$$

Step 4. Checking of the final logic circuit device, Fig. 6.24.

Example 2.
Design logic circuit on the logic gates NAND or NOR.

$$Y = \vee(1,4,6,10,11,12,12,13,14) \quad \text{or} \quad Y = \wedge(0,2,3,5,7,8,9,15).$$

First variant (logic circuit on the logic gates NAND) (Fig. 6.25).

$$y = \overline{X}_3\overline{X}_2\overline{X}_1X_0 + X_3\overline{X}_2X_1 + X_3X_2\overline{X}_1 + X_2\overline{X}_0.$$

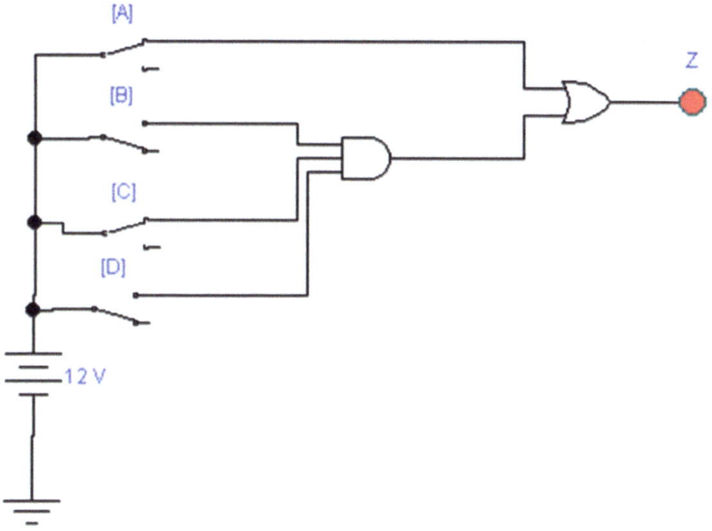

Fig. 6.24

	$\overline{X}_1\overline{X}_0$	$\overline{X}_1X_0$	X_1X_0	$X_1\overline{X}_0$
$\overline{X}_3\overline{X}_2$	0	1	0	0
$\overline{X}_3X_2$	1	0	0	1
X_3X_2	1	1	0	1
$X_3\overline{X}_2$	0	0	1	1

$y = \overline{X}_3\overline{X}_2\overline{X}_1X_0 +$
$+ X_3\overline{X}_2X_1 + X_3X_2\overline{X}_1 +$
$+ X_2\overline{X}_0.$

Fig. 6.25 Minimization of the function with gates NAND

Since $\overline{X} = \overline{X} + \overline{X} = \overline{X \bullet Y}$

$$y = \overline{X}_3\overline{X}_2\overline{X}_1X_0 + X_3\overline{X}_2X_1 + X_3X_2\overline{X}_1 + X_2\overline{X}_0 =$$

$$= \overline{\overline{\overline{\overline{X}_3\overline{X}_2\overline{X}_1X_0} \bullet \overline{X_3X_2\overline{X}_1} \bullet \overline{X_3\overline{X}_2X_1} \bullet \overline{X_2\overline{X}_0}}} =$$

$$= \overline{\overline{\overline{X_2 \bullet \overline{\overline{X_0X_0}}} \bullet \overline{X_3 \bullet X_2 \bullet \overline{\overline{X_1X_1}}} \bullet \overline{X_3 \bullet X_1 \bullet \overline{\overline{X_2X_2}}} \bullet \overline{\overline{X_3 \bullet X_3}} \bullet \overline{\overline{X_2X_2}} \bullet \overline{\overline{X_1X_1}} \bullet X_0}}.$$

Let

$$A = X_0;$$
$$B = X_1;$$
$$C = X_2;$$
$$D = X_3.$$

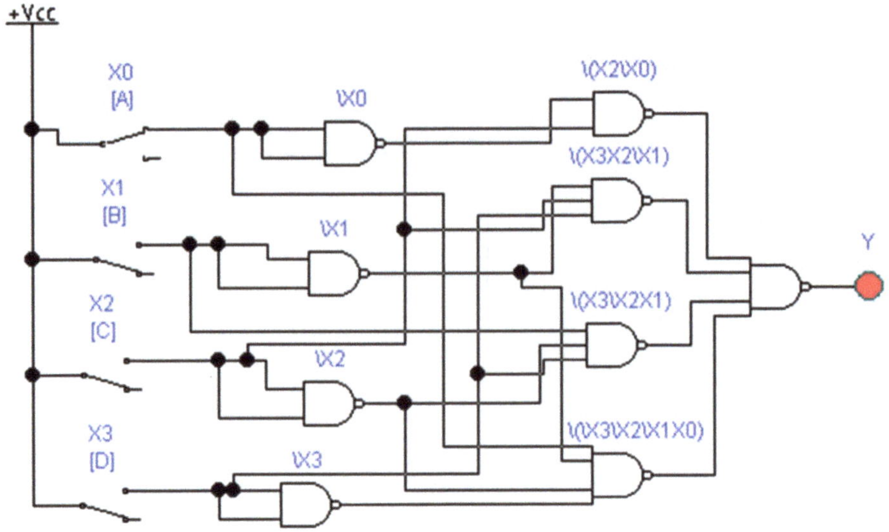

Fig. 6.26 Checking of the final logic circuit with gates NAND

	$\bar{X}_1\bar{X}_0$	$\bar{X}_1X_0$	X_1X_0	$X_1\bar{X}_0$
$\bar{X}_3\bar{X}_2$	0	1	0	0
$\bar{X}_3X_2$	1	0	0	1
X_3X_2	1	1	0	1
$X_3\bar{X}_2$	0	0	1	1

$$y = (X_3 + X_2 + X_0)\cdot$$
$$\cdot(X_3 + X_2 + \bar{X}_1)\cdot$$
$$\cdot(\bar{X}_3 + X_2 + X_1)\cdot$$
$$\cdot(X_3 + \bar{X}_2 + \bar{X}_0)\cdot$$
$$\cdot(\bar{X}_2 + \bar{X}_1 + \bar{X}_0).$$

Fig. 6.27 Minimization of the function with gates NOR

Checking of the final logic circuit, Fig. 6.26.
Second variant (logic circuit on the logic gates NOR) (Fig. 6.27).

$$y = (X_3 + X_2 + X_1)\cdot(X_3 + X_2 + \bar{X}_1)\cdot(\bar{X}_3 + X_2 + X_1)\cdot$$
$$(X_3 + \bar{X}_2 + \bar{X}_0)\cdot(\bar{X}_2 + \bar{X}_1 + \bar{X}_0)$$
$$= \overline{\overline{X_3 + X_2 + X_1} + \overline{X_3 + X_2 + \bar{X}_1} + \overline{\bar{X}_3 + X_2 + X_1} + \overline{X_3 + \bar{X}_2 + \bar{X}_0} + \overline{\bar{X}_2 + \bar{X}_1 + \bar{X}_0}}.$$

Checking of the final logic circuit, Fig. 6.28.
Let

$$A = X_0;\quad B = X_1;\quad C = X_2;\quad D = X_3.$$

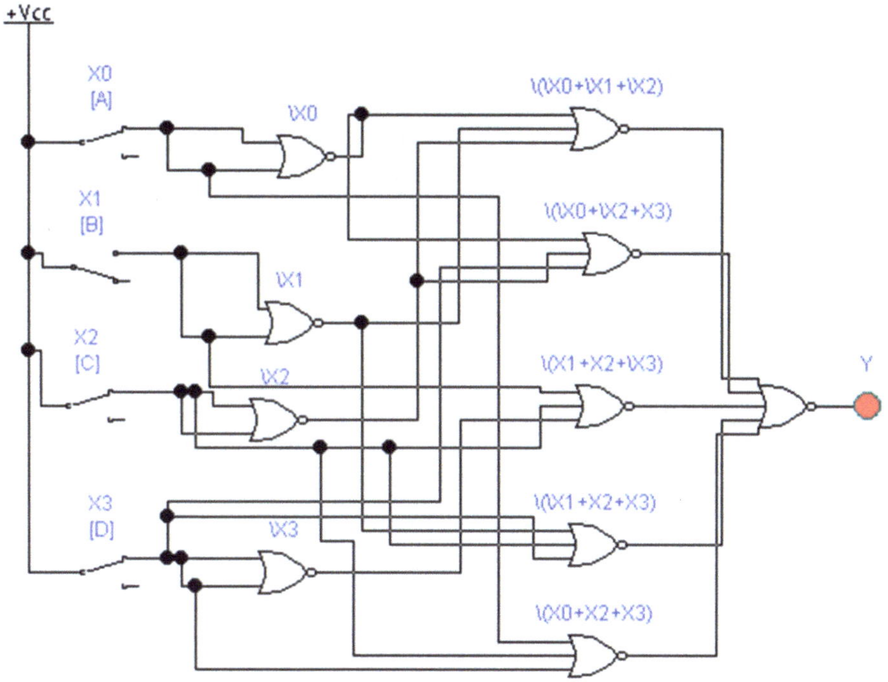

Fig. 6.28 Checking of the final logic circuit with gates NOR

6.7 Evaluating Logic Circuit Outputs

In general, the following rules must always be followed when evaluating a Boolean expression:

1. First, perform all inversions of single terms; that is, $\overline{0}=1$ or $\overline{1}=0$.
2. Then perform all operations within parentheses.
3. Perform an AND operation before an OR operation unless parentheses indicate otherwise.
4. If an expression has a bar over it, perform the operations of the expression first and then invert the result.

Examples

Given the following Boolean variables *A* = *0*, *B* = *1*, *C* = *1*, *D* = *1*. Find *Y* (Fig. 6.29).

Solution:

$$Y=\overline{A}\cdot B\cdot C\cdot(\overline{A+D})=\overline{0}\cdot 1\cdot 1\cdot(\overline{0+1})=1\cdot 1\cdot 1\cdot(\overline{0+1})=1\cdot 1\cdot 1\cdot(\overline{1})=1\cdot 1\cdot 1\cdot(0)=0.$$

Give.n the following Boolean variables *A* = *0*, *B* = *1*, *C* = *1*, *D* = *1*. Find *Y* (Fig. 6.30).

Solution:

$$Y=\overline{A}\cdot B\cdot C+(\overline{A+D})=\overline{0}\cdot 1\cdot 1+(\overline{0+1})=1\cdot 1\cdot 1+(\overline{0+1})=1\cdot 1\cdot 1+(\overline{1})=1\cdot 1\cdot 1+(0)=1.$$

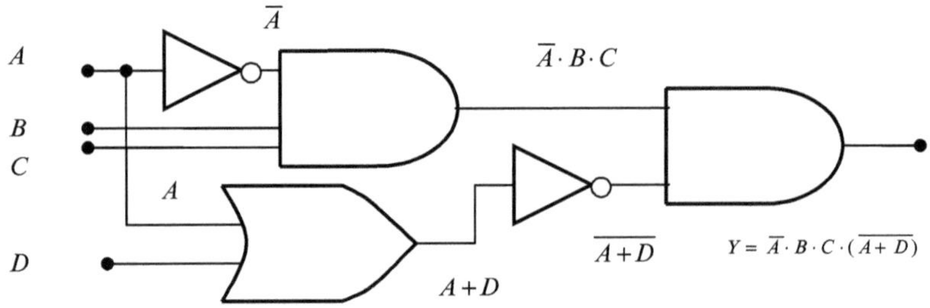

Fig. 6.29

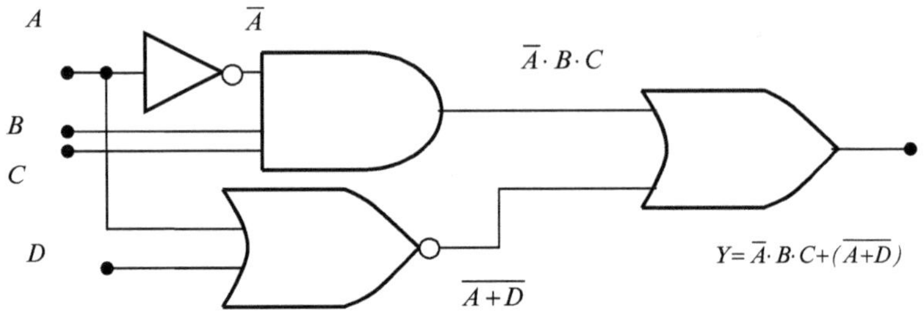

Fig. 6.30

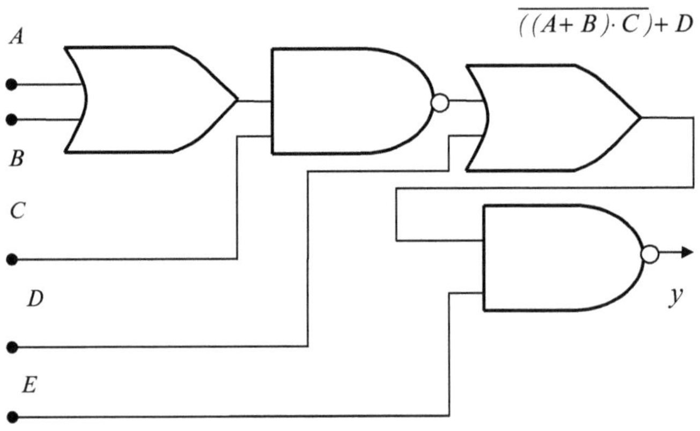

Fig. 6.31

Given the following Boolean variables $A=0,\ B=1,\ C=1,\ D=1,\ E=1$. Find Y (Fig. 6.31).

Solution:

$$Y=\overline{(\overline{\overline{((A+B)\cdot C)}}+D)\cdot E}=\overline{(\overline{\overline{((0+1)\cdot 1)}}+1)\cdot 1}=\overline{(\overline{\overline{0\cdot 1+1\cdot 1)}}+1)\cdot 1}=\overline{(\overline{\overline{(0+1)}}+1)\cdot 1}$$

$$=\overline{(\overline{\overline{1}})+1)\cdot 1}=\overline{(0+1)\cdot 1}=\overline{(1)\cdot 1}=\overline{1}=0.$$

By combination circuits we mean logical circuits which do not contain feedback.

Existing methods for the synthesis of combination circuits cover only the first part of the problem, the construction and minimization of logical control in Boolean operations.

The other steps of synthesis which are essential to electronic circuits have not been formalised. These include:

1. The expression of Boolean equations in a given operator system.
2. Attaining the desired quality for the physical characteristics of the network.
3. The comparison of different versions of the network.

The existence of efficient algorithms for these steps simplifies the synthesis of actual electronic circuits satisfying given reliability criteria with the minimum use of equipment. There is also the possibility of the complete automation of the synthesis with the help of digital computers.

Chapter 7
Minimizing Boolean Functions

Abstract This chapter describes the graphical and algebraic most widely used ways to minimize logic functions (in order to reduce the circuit's complexity), like truth tables, Karnaugh Maps that are based on the rule of complementation and the Quine-Mccluskey method, which is functionally identical to Karnaugh mapping, but its tabular form makes it more efficient for use in computer algorithms. The chapter provides many examples of minimization and their hardware implementations.

7.1 Background and Terminology

Engineers use many methods to minimize logic functions, in order to reduce the circuit's complexity. When the complexity is less, the circuit also has fewer errors and less electronics, and is therefore less expensive.

The most widely used simplification is a minimization algorithm like the Espresso heuristic logic minimizer within a CAD system, although historically, binary decision diagrams, an automated Quine-McCluskey algorithm, truth tables, Karnaugh Maps, and Boolean algebra have been used.

This chapter describes graphical and algebraic ways to minimize Boolean functions.

All the data path and control structures of a digital device can be represented as Boolean functions, which take the general form, Fig. 7.1.:

$$
\begin{aligned}
Y_1 &= F_1(X_1, X_2, \ldots, X_m); \\
Y_2 &= F_2(X_1, X_2, \ldots, X_m); \\
&\ldots\ldots\ldots\ldots\ldots\ldots\ldots \\
Y_n &= F_n(X_1, X_2, \ldots, X_m),
\end{aligned}
$$

B. Borowik et al., *Theory of Digital Automata*, Intelligent Systems, Control and Automation: Science and Engineering 63, DOI 10.1007/978-94-007-5228-3_7,

Fig. 7.1 The structure of a digital device

where X_1, X_2, ..., X_m – is a set of Boolean variables (variables that may take on only the values zero and one). These Boolean functions must be converted into logic networks in the most economical way possible. What qualifies as the "most economical way possible" varies, depending on whether the network is built using discrete gates, a programmable logic device with a fixed complement of gates available, or a fully-customized integrated circuit. But in all cases, minimization yields a network with as small a number of gates as possible, and with each gate as simple as possible.

The variables in the expression on the right side of a Boolean equation are the input wires to a logic network. The left side of a Boolean equation is the output wire of the network.

Any Boolean equation or combinational logic network can be completely and exactly characterized by a truth table. A truth table lists every possible combination of values of the input variables, and the corresponding output value of the function for each combination. There are 2^n rows in the truth table for a function or network with n input variables, so it isn't always practical to write out an entire truth table. But for relatively small values of n, a truth table provides a convenient way to describe the function or network's behavior exactly.

Every row of a truth table with a one in the output column is called a minterm. A convenient way to represent a truth table is to treat each combination of input variables as a binary number and to list the numbers of the rows that are minterms.

Representations are crucial to an engineer's design of digital circuits. Some analysis methods only work with particular representations.

The classical way to represent a digital circuit is with an equivalent set of logic gates. Another way, often with the least electronics, is to construct an equivalent system of electronic switches (usually transistors). One of the easiest ways is to simply have a memory containing a truth table. The inputs are fed into the address of the memory, and the data outputs of the memory become the outputs.

For automated analysis, these representations have digital file formats that can be processed by computer programs. Most digital engineers are very careful to select computer programs ("tools") with compatible file formats [13, 14].

This document uses the function with the following truth table as a running example (Table 7.1):

This truth table can also be represented as the list of minterms, [10, 11, 12, 13, 14, 15, 16, 17, 18]. That is, the truth table has a 1 in the Y column for the rows where the binary number represented by the values of A, B, and C is one of the numbers listed inside the square brackets. The other two rows (0, 1, 2 and 6) have a 0 in the Y column, and thus are not minterms.

One standard way to represent any Boolean function is called "sum of products" (SOP) or, more formally, disjunctive normal form. In this form, the function is written as the logical **OR** (indicated by +) of a set of **AND** terms, one per minterm.

Table 7.1 Truth table

№ minterm	A	B	C	Y
0	0	0	0	0
1	0	0	1	0
2	0	1	0	0
3	0	1	1	1
4	1	0	0	1
5	1	0	1	1
6	1	1	0	0
7	1	1	1	1

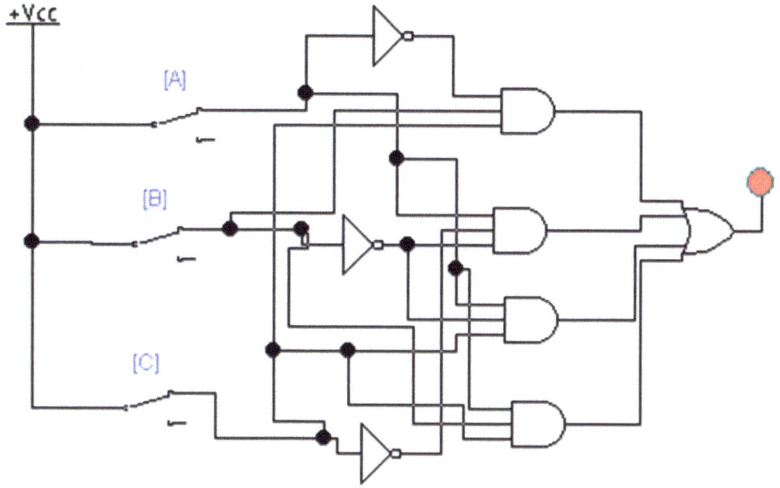

Fig. 7.2 The logic circuit

For example, the disjunctive normal form for our sample function would be:

$$Y(A,B,C) = \overline{A}{\cdot}B{\cdot}C + A{\cdot}\overline{B}{\cdot}\overline{C} + A{\cdot}\overline{B}{\cdot}C + ABC.$$

There is also a conjunctive normal form, which represents an expression as a product of sums rather than as a sum of products. The material presented below can be extended to deal with conjunctive normal forms rather than disjunctive normal forms.

A literal is a variable that is either complemented or not in a product term. The minterms in our sample function have a total of six literals: $\overline{A}, A, \overline{B}, B, \overline{C}, C$.

To appreciate the importance of minimization, consider the two networks in Figs. 7.2 and 7.3. Both behave exactly the same way. No matter what pattern of ones and zeros you put into a, b, and c in Fig. 7.2, the value it produces at y will be exactly matched if you put the same pattern of values into the corresponding inputs in Fig. 7.3. Yet the network in Fig. 7.3 uses far fewer gates, and the gates it uses are

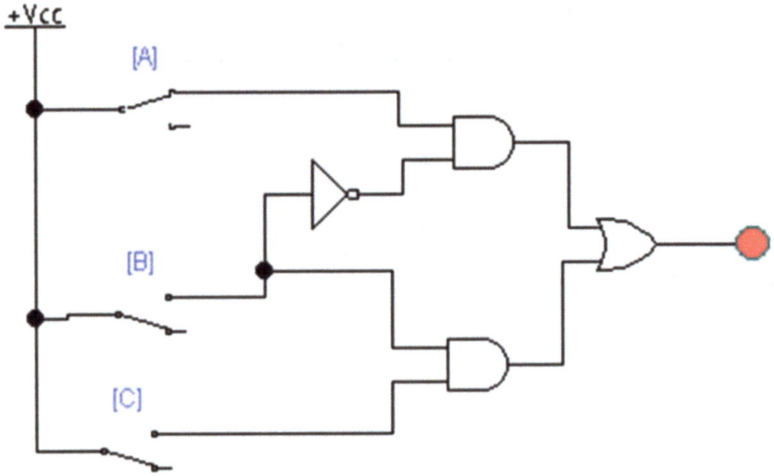

Fig. 7.3 The logic circuit

simpler (have smaller fan-ins) than the gates in Fig. 7.2. Clearly, the minimized circuit should be less expensive to build than the unminimized version. Although it is not true for Fig. 7.3, it is often the case that minimized networks will be faster (have fewer propagation delays) than unminimized networks.

The network in Fig. 7.3 uses only 4 literals because $\overline{A},\overline{C}$ isn't used. In the disjunctive normal form of a function, each product term has one literal for each variable.

$$Y(A,B,C) = \overline{A}{\cdot}B{\cdot}C + A{\cdot}\overline{B}{\cdot}\overline{C} + A{\cdot}\overline{B}{\cdot}C + ABC = A\overline{B} + BC$$

Figure 7.2 implements our sample function, and demonstrates translating a disjunctive normal form function directly into a logic network.

There are many rules for manipulating a Boolean expression algebraically, but there is just one rule that you need in order to minimize a function once it is in disjunctive normal form: the rule of complementation.

For example:

$$\begin{aligned}
y = f(A,B,C) &= (\overline{A}{\cdot}\overline{B}{\cdot}C) \vee (A{\cdot}\overline{B}{\cdot}\overline{C}) \vee (A{\cdot}\overline{B}{\cdot}C) \vee (A{\cdot}B{\cdot}\overline{C}) \vee (A{\cdot}B{\cdot}C) = \\
&= (\overline{A}{\cdot}\overline{B}{\cdot}C \vee A{\cdot}\overline{B}{\cdot}C) \vee (A{\cdot}\overline{B}{\cdot}\overline{C} \vee A{\cdot}\overline{B}{\cdot}C) \vee (A{\cdot}B{\cdot}\overline{C} \vee A{\cdot}B{\cdot}C) = \\
&= \overline{B}{\cdot}C{\cdot}(A \vee \overline{A}) \vee A{\cdot}\overline{B}{\cdot}(\overline{C} \vee C) \vee A{\cdot}B{\cdot}(C \vee \overline{C}) = \overline{B}{\cdot}C \vee A{\cdot}\overline{B} \vee A{\cdot}B = \\
&= \overline{B}{\cdot}C \vee A{\cdot}(\overline{B} \vee B) = \overline{B}{\cdot}C \vee A.
\end{aligned}$$

The rule of complementation says that $(\overline{A} \vee A)$ is always true (1), so any two terms that are in the form $(\overline{A} \vee A){\cdot}B$ can be reduced to just B without changing its meaning. Another way of saying this is that two product terms can be simplified if the only difference between them is the value of exactly one variable, in which case that variable can be eliminated from both terms to give an equivalent single term.

Table 7.2 Minimizing Boolean functions

Boolean expression	Clarification
$y=\bar{A}\cdot\bar{B}+A\cdot\bar{B}+A\cdot B+B\cdot C=$	
$=\bar{A}\cdot\bar{B}+A\cdot\bar{B}+A\cdot\bar{B}+A\cdot B+B\cdot C=$	$x\vee x=x$
$=\left(\bar{B}\cdot\bar{A}+\bar{B}\cdot A+A\cdot\bar{B}+A\cdot B+B\cdot C=\right)$	$x\cdot y=y\cdot x$
$=\bar{B}\cdot(\bar{A}+A)+A\cdot(\bar{B}+B)+B\cdot C=$	$x\cdot(y+z)=x\cdot y+x\cdot z$
$=\bar{B}\cdot(A+\bar{A})+A\cdot(B+\bar{B})+B\cdot C=$	$x+y=y+x$
$=\bar{B}\cdot 1+A\cdot 1+B\cdot C=$	$x+\bar{x}=1$
$=\bar{B}+A+B\cdot C=$	$x\cdot 1=x$
$=A+\bar{B}+B\cdot C=$	$x+y=y+x$
$=A+\bar{B}+C$	$x+\bar{x}\cdot y=x+y$ or $\bar{x}+x\cdot y=\bar{x}+y$

Table 7.3 Minimizing Boolean functions

Boolean expression	Clarification
$y=(A+\bar{B}+C)\cdot(\overline{A+B+C})$	
$=(A+\bar{B}+C)\cdot\bar{A}\cdot\bar{B}\cdot\bar{C}=$	$\overline{x+y+z}=\bar{x}\cdot\bar{y}\cdot\bar{z}$
$=A\cdot\bar{A}\cdot\bar{B}\cdot\bar{C}+\bar{A}\cdot\bar{B}\cdot\bar{B}\cdot\bar{C}+\bar{A}\cdot\bar{B}\cdot\bar{C}\cdot C=$	
$=\bar{A}\cdot\bar{B}\cdot\bar{C}$	$x\cdot\bar{x}=0$ and $\bar{x}\cdot\bar{x}=\bar{x}$

Table 7.4 Minimizing Boolean functions

Boolean expression	Clarification
$y=A\cdot(\bar{B}\cdot\bar{D}+C)+\overline{\bar{A}\cdot\bar{B}\cdot\bar{D}+A\bar{C}}+\overline{A\cdot(\bar{B}\cdot\bar{D}+\bar{C})}=$	
$=A\cdot\bar{B}\cdot\bar{D}+A\cdot C+\overline{(\bar{A}\cdot\bar{B}\cdot\bar{D})\cdot(\overline{A\bar{C}})}+\overline{A\cdot\bar{B}\cdot\bar{D}+A\cdot\bar{C}}=$	$\overline{x+y+z}=\bar{x}\cdot\bar{y}\cdot\bar{z}$
$=A\cdot\bar{B}\cdot\bar{D}+A\cdot C+(A+B+D)(\bar{A}+C)+\overline{(A\cdot\bar{B}\cdot\bar{D})}\cdot\overline{(A\cdot\bar{C})}=$	$\bar{x}\cdot\bar{y}\cdot\bar{z}=\overline{x+y+z}$
$=A\cdot\bar{B}\cdot\bar{D}+A\cdot C+A\cdot C+\bar{A}\cdot B+B\cdot C+\bar{A}\cdot D$ $+C\cdot D+(\bar{A}+B+D)\cdot(\bar{A}+C)=$	
$=A\cdot\bar{B}\cdot\bar{D}+A\cdot C+A\cdot C+\bar{A}\cdot B+B\cdot C+\bar{A}\cdot D$ $+C\cdot D+(\bar{A}+B+D)\cdot(\bar{A}+C)=$	
$=A\cdot\bar{B}\cdot\bar{D}+A\cdot C+A\cdot C+\bar{A}\cdot B+B\cdot C+\bar{A}\cdot D+C\cdot D$ $+\bar{A}+\bar{A}\cdot B+\bar{A}\cdot C+B\cdot C+\bar{A}\cdot D+C\cdot D=$	
$=A\cdot\bar{B}\cdot\bar{D}+C\cdot(A+\bar{A})+\bar{A}\cdot(B+1)+B\cdot C+\bar{A}\cdot(1+D)+C\cdot D=$	$x+\bar{x}=1$
$=A\cdot\bar{B}\cdot\bar{D}+C+\bar{A}=$	$x+\bar{x}\cdot y=x+y$ or $\bar{x}+x\cdot y=\bar{x}+y$
$=\bar{B}\cdot\bar{D}+C+\bar{A}$	

For example $A\cdot\bar{B}\cdot C+\bar{A}\cdot\bar{B}\cdot C$ is equivalent to $(A+\bar{A})\cdot\bar{B}\cdot C$, which is the same as the single product term, $\bar{B}\cdot C$.

Consider a few examples (Tables 7.2, 7.3, 7.4, and 7.5).

Example

Thus, we obtain $y=\bar{A}\cdot\bar{B}+A\cdot\bar{B}+A\cdot B+B\cdot C=A+\bar{B}+C$.

Table 7.5 Minimizing Boolean functions

Boolean expression	Clarification
$y = A\cdot\overline{B}\cdot\overline{D} + C\cdot\overline{D} + \overline{A\cdot\overline{B}\cdot\overline{C} + \overline{A}\cdot B\cdot\overline{C}} + \overline{D} + \overline{A\cdot(\overline{B}+\overline{C})} =$	
$= A\cdot\overline{B}\cdot\overline{D} + C\cdot\overline{D} + (\overline{A}+B+C)\cdot(A+\overline{B}+C) + \overline{D} + \overline{A} + B\cdot C =$	$\overline{x+y+z} = \overline{x}\cdot\overline{y}\cdot\overline{z}$
$= \overline{D} + C\cdot\overline{D} + \overline{A}\cdot\overline{B} + B\cdot A + \overline{A} =$	
$= \overline{D} + C + \overline{A} + B$	

Example

Thus, we obtain $y = (A+\overline{B}+C)\cdot\overline{(A+B+C)} = \overline{A}\cdot\overline{B}\cdot\overline{C}$

Example

Thus, we obtain

$$y = A\cdot(\overline{B}\cdot\overline{D}+C) + \overline{\overline{A}\cdot\overline{B}\cdot\overline{D} + A\overline{C}} + \overline{A\cdot(\overline{B}\cdot\overline{D}+\overline{C})} = \overline{B}\cdot\overline{D} + C + \overline{A}.$$

Example

Thus, we obtain

$$y = A\cdot\overline{B}\cdot\overline{D} + C\cdot\overline{D} + \overline{A\cdot\overline{B}\cdot\overline{C} + \overline{A}\cdot B\cdot\overline{C}} + \overline{D} + \overline{A\cdot(\overline{B}+\overline{C})} = \overline{D} + C + \overline{A} + B.$$

7.2 Karnaugh Maps

A Karnaugh Map is a graphical way of minimizing a Boolean expression based on the rule of complementation. It works well if there are 2, 3, or 4 variables, but gets messy or impossible to use for expressions with more variables than that.

The idea behind a Karnaugh Map (Karnaugh 1953) is to draw an expression's truth table as a matrix in such a way that each row and each column of the matrix puts minterms that differ in the value of a single variable adjacent to each other. Then, by grouping adjacent cells of the matrix, you can identify product terms that eliminate all complemented literals, resulting in a minimized version of the expression [4, 5, 19].

The Karnaugh map (K-map for short), Maurice Karnaugh's refinement of Edward Veitch's (1952) Veitch diagram, is a method to simplify Boolean algebra expressions. The Karnaugh map reduces the need for extensive calculations by taking advantage of humans' pattern-recognition capability, permitting the rapid identification and elimination of potential race conditions (Table 7.6).

Truth table

Figure 7.4 shows how the minterms in truth Table 7.4 are placed in a Karnaugh Map grid for both 2-variable expressions.

Table 7.6 Truth table

A	*B*	*f(A,B)*
0	0	*f*(0,0)
0	1	*f*(0,1)
1	0	*f*(1,0)
1	1	*f*(1,1)

Fig. 7.4 Karnaugh Map

	B = 0	*B = 1*
A = 0	*f*(0,0)	*f*(0,1)
A = 1	*f*(1,0)	*f*(1,1)

Truth table			
A	*B*	*C*	*f(A,B,C)*
0	**0**	**0**	*f*(0,0,0)
0	**0**	**1**	*f*(0,0,1)
0	**1**	**0**	*f*(0,1,0)
0	**1**	**1**	*f*(0,1,1)
1	**0**	**0**	*f*(1,0,0)
1	**0**	**1**	*f*(1,0,1)
1	**1**	**0**	*f*(1,1,0)
1	**1**	**1**	*f*(1,1,1)

Karnaugh Map grid for both 3-variable expressions				
	BC			
	00	**01**	**11**	**10**
A	*f*(0,00)	*f*(0,01)	*f*(0,11)	*f*(0,10)
	f(1,00)	*f*(1,01)	*f*(1,11)	*f*(1,10)

Fig. 7.5 The minterms in truth tables and Karnaugh Map

f(A,B,C,D)		*CD*			
		00	**01**	**11**	**10**
AB	**00**	*f*(0,0,0,0)	*f*(0,0,0,1)	*f*(0,0,1,1)	*f*(0,0,1,0)
	01	*f*(0,1,0,0)	*f*(0,1,0,1)	*f*(0,1,1,1)	*f*(0,1,1,0)
	11	*f*(1,1,0,0)	*f*(1,1,0,1)	*f*(1,1,1,1)	*f*(1,1,1,0)
	10	*f*(1,0,0,0)	*f*(1,0,0,1)	*f*(1,0,1,1)	*f*(1,0,1,0)

Fig. 7.6 Karnaugh Map

Figures 7.5 and 7.6 shows how the minterms in truth tables are placed in a Karnaugh Map grid for both 3 and 4-variable expressions.

Figures 7.7, 7.8, 7.9, and 7.10 shows how the minterms in truth tables are placed in a Veitch diagram for both 2, 3, 4 and 5-variable expressions.

The input variables can be combined in 16 different ways, so the Karnaugh map has 16 positions, and therefore is arranged in a 4×4 grid.

Fig. 7.7 The Veitch diagram for 2 variables

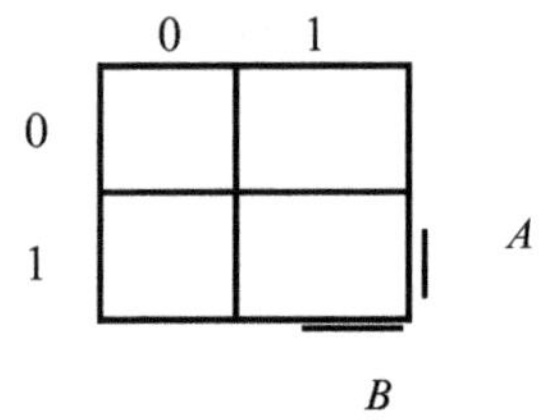

Fig. 7.8 The Veitch diagram for 3 variables

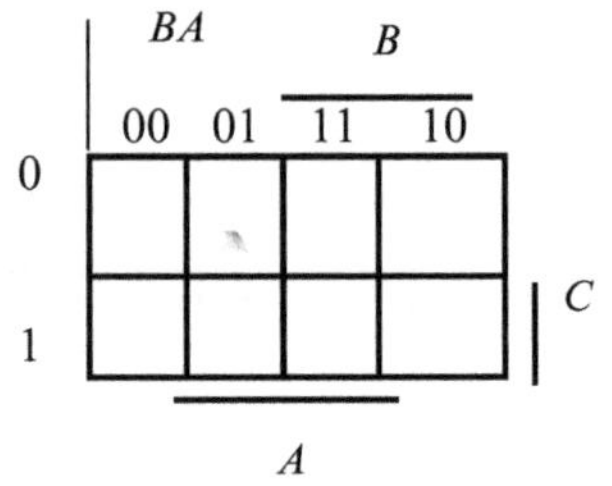

Fig. 7.9 The Veitch diagram for 4 variables

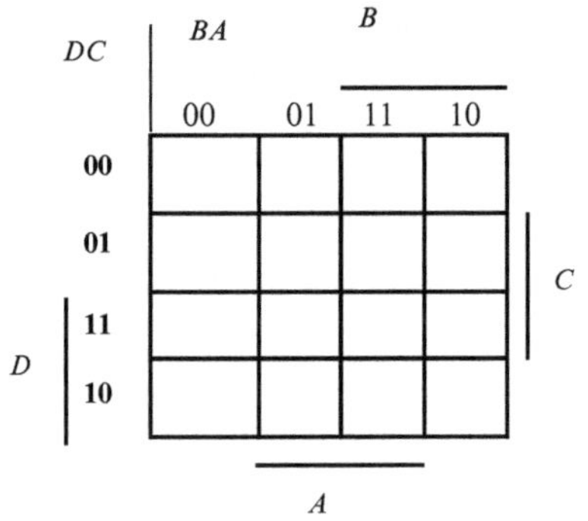

Fig. 7.10 The Veitch diagram for 5 variables

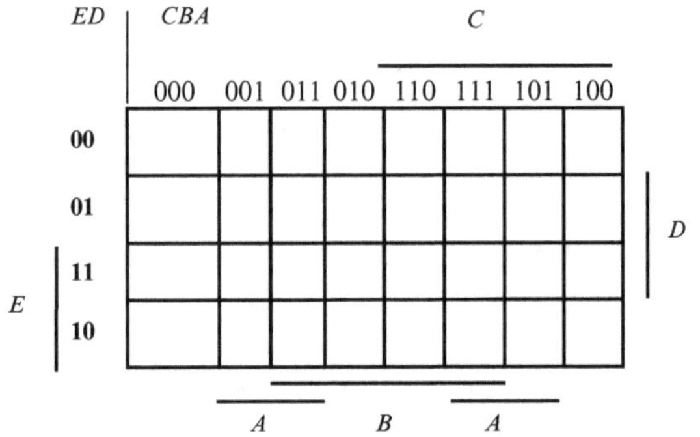

Fig. 7.11 Truth tables and Karnaugh Map

Truth table

A	*B*	*C*	*f(A,B,C)*
0	0	0	1
0	0	1	1
0	1	0	1
0	1	1	1
1	0	0	0
1	0	1	1
1	1	0	0
1	1	1	0

Karnaugh Map grid for both 3-variable expressions

A \ *BC*	00	01	11	10
0	1	1	1	1
1	0	1	0	0

The binary digits in the map represent the function's output for any given combination of inputs. So 0 is written in the upper leftmost corner of the map because $f = 0$ when $A = 0,\ B = 0,\ C = 0,\ D = 0$. Similarly we mark the bottom right corner as 1 because $A = 1,\ B = 0,\ C = 1,\ D = 0$ gives $f = 1$. Note that the values are ordered in a Gray code, so that precisely one variable changes between any pair of adjacent cells.

After the Karnaugh map has been constructed the next task is to find the minimal terms to use in the final expression. These terms are found by encircling groups of 1's in the map. The groups must be rectangular and must have an area that is a power of two (i.e. $2^n = 1, 2, 4, 8\ldots$). The rectangles should be as large as possible without containing any 0's.

Figure 7.11 shows example how the minterms in truth tables are placed in a Karnaugh Map grid for both 3-variable expressions.

$$f(A,B,C) = \overline{A}{\cdot}\overline{B}{\cdot}\overline{C} \vee \overline{A}{\cdot}\overline{B}{\cdot}C \vee \overline{A}{\cdot}B{\cdot}\overline{C} \vee \overline{A}{\cdot}B{\cdot}C \vee A{\cdot}\overline{B}{\cdot}C == \overline{A} \vee \overline{B}{\cdot}C.$$

Example 1

$$\begin{aligned} Y = {} & \overline{D}\cdot\overline{C}\overset{1}{\cdot}B\cdot A + \overline{D}\cdot C\overset{2}{\cdot}\overline{B}\cdot\overline{A} + \overline{D}\cdot C\overset{3}{\cdot}B\cdot\overline{A} \\ & + \overline{D}\cdot C\overset{4}{\cdot}B\cdot A + D\cdot\overline{C}\overset{5}{\cdot}B\cdot\overline{A} + D\cdot\overline{C}\overset{6}{\cdot}B\cdot A \\ & + D\cdot C\overset{7}{\cdot}\overline{B}\cdot\overline{A} + D\cdot C\cdot B\cdot\overline{A}. \end{aligned}$$

A Karnaugh Map (Veitch diagram) is used to produce a minimal sum of products implementation of an expression by drawing rectangles around groups of adjacent minterms in the map; each rectangle will correspond to one product term, and the full expression will be constructed as the OR (sum) of all the product terms. The goal is to have as few product terms as possible, which implies that each product term will account for as many minterms as possible (Fig. 7.12).

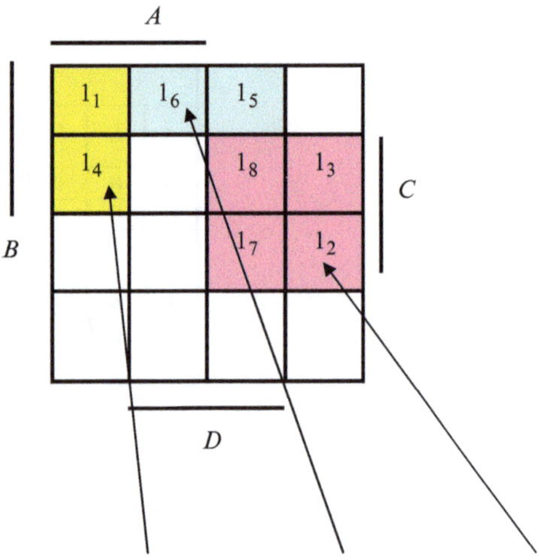

Fig. 7.12 The Veitch diagram for 4 variables

$$Y = \overline{D} \cdot B \cdot A + D \cdot \overline{C} \cdot B + C \cdot \overline{A}.$$

Here are the rules for drawing the rectangles:

Every minterm must be inside at least one rectangle, but there must not be any zeros inside any rectangles.
Every rectangle has to be as large as possible.
Rectangles may wrap around to include cells in both the leftmost and rightmost columns, likewise for the top and bottom rows.
The number of minterms enclosed in a rectangle must be a power of 2 (1, 2, 4, 8, or 16 minterms for 4-variable maps).

Some functions have "don't care" conditions, which are combinations of inputs that will never occur, resulting in cases where it doesn't matter whether the output is a 0 or a 1. Where these "don't care" conditions appear in a Karnaugh Map (usually indicated by X's instead of 1's or 0's), they may be included inside rectangles or not depending on what will make the rectangles as few and as large as possible.

$$Y = \overline{D}{\cdot}B{\cdot}A + D{\cdot}\overline{C}{\cdot}B + C{\cdot}\overline{A}.$$

Example 2

Truth table (Table 7.7)

"Don't cares" in a Karnaugh map, or truth table, may be either 1's or 0's, as long as we don't care what the output is for an input condition we never expect to see. We plot these cells with an asterisk, *, among the normal 1's and 0's. When forming groups of cells, treat the "don't care" cell as either a 1 or a 0, or ignore the "don't cares". This is helpful if it allows us to form a larger group than would otherwise be possible without the "don't cares". There is no requirement to group all or any of the "don't cares". Only use them in a group if it simplifies the logic.

Table 7.7 Truth table

	D	A	B	C	Z	
0	0	0	0	0	**0**	
1	0	0	0	1	**1**	$\overline{A}\cdot\overline{B}\cdot C\cdot\overline{D}=1$
2	0	0	1	0	**1**	$\overline{A}\cdot B\cdot\overline{C}\cdot\overline{D}=1$
3	0	0	1	1	*	
4	0	1	0	0	**1**	$A\cdot\overline{B}\cdot\overline{C}\cdot\overline{D}=1$
5	0	1	0	1	*	
6	0	1	1	0	*	
7	0	1	1	1	*	
8	1	0	0	0	**0**	
9	1	0	0	1	**0**	
10	1	0	1	0	**0**	
11	1	0	1	1	*	
12	1	1	0	0	**0**	
13	1	1	0	1	*	
14	1	1	1	0	*	
15	1	1	1	1	*	

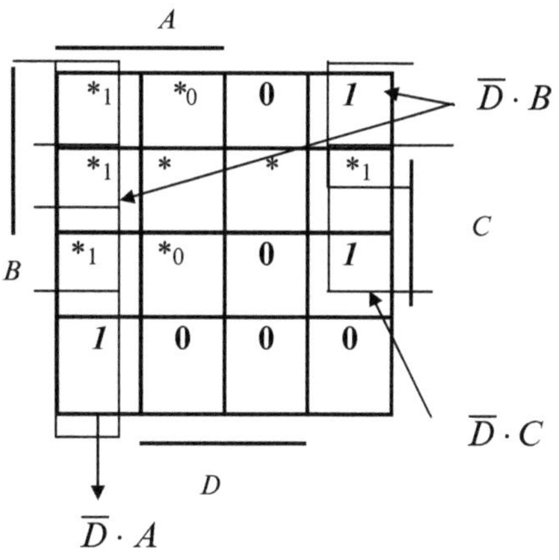

Fig. 7.13 The Veitch diagram for 4 variables

$$Z = A\cdot\overline{B}\cdot\overline{C}\cdot\overline{D} + \overline{A}\cdot\overline{B}\cdot C\cdot\overline{D} + \overline{A}\cdot B\cdot\overline{C}\cdot\overline{D}$$

Figures 7.13 and 7.14 show an example of a Veitch diagram and Karnaugh Map for both 4-variable expressions.

$$Z = \overline{D}\cdot A + \overline{D}\cdot B + \overline{D}\cdot C = \overline{D}(A + B + C).$$

$$Z = \overline{D}(A + B + C) = \overline{D}\cdot A + \overline{D}\cdot B + \overline{D}\cdot C.$$

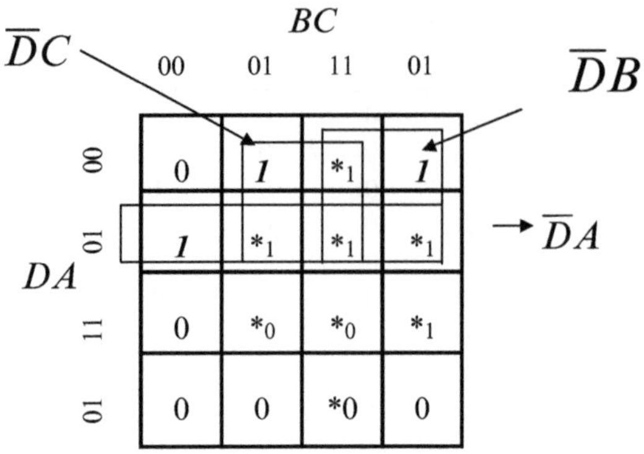

Fig. 7.14 The Karnaugh Map for 4 variables

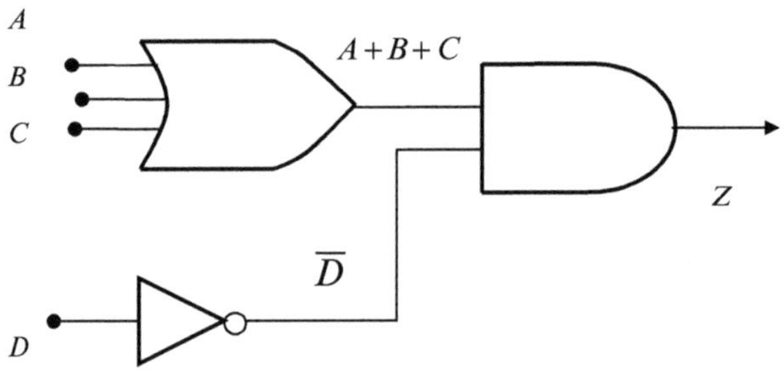

Fig. 7.15 The network

$$Z=\overline{D}(A+B+C)=\overline{D}\cdot A+\overline{D}\cdot B+\overline{D}\cdot C.$$

$$Z=\overline{(\overline{\overline{\overline{A}\cdot\overline{B}\cdot\overline{C}}})+(\overline{\overline{D}})}$$

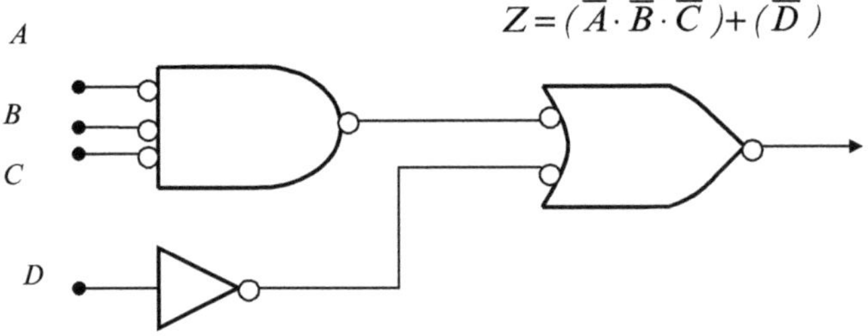

Fig. 7.16 The network after optimization

The network in Fig. 7.15 uses only 4 literals because $\overline{A},\overline{C}$ isn't used

After optimization of network we will get a next expression, Fig. 7.16.

Figure 7.17 shows the networks collected in Multisim 10.

The set of all essential prime implicants must be contained in any irredundant sum-of-products expression, while any prime implicant covered by the sum of the essential prime implicants must not be contained in an irredundant expression. For example, the prime implicant $B{\cdot}D$ of function y of Fig. 7.18 is covered by the sum

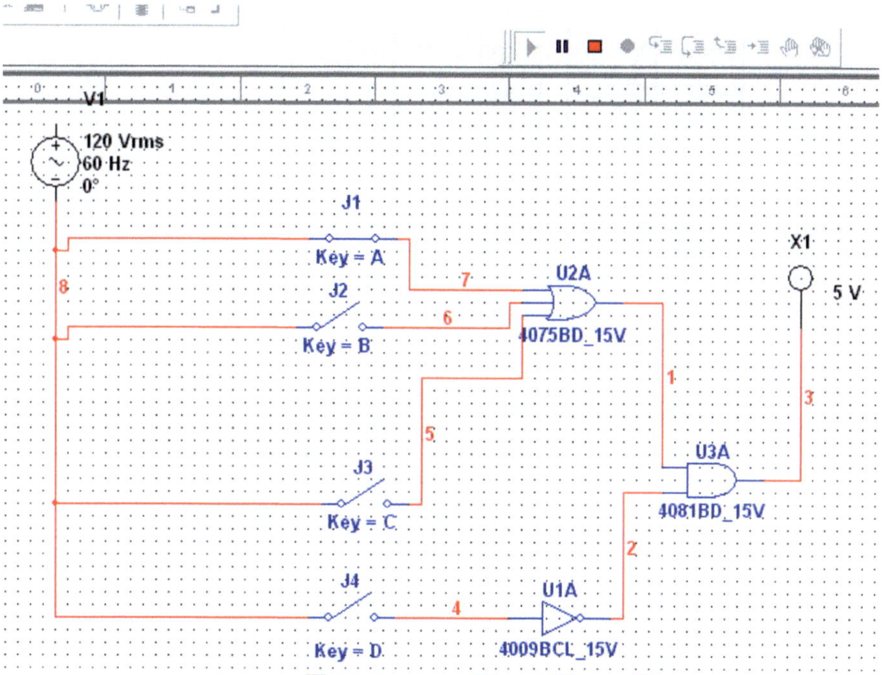

a) $Z=\overline{D}(A+B+C)=\overline{D}\cdot A+\overline{D}\cdot B+\overline{D}\cdot C.$

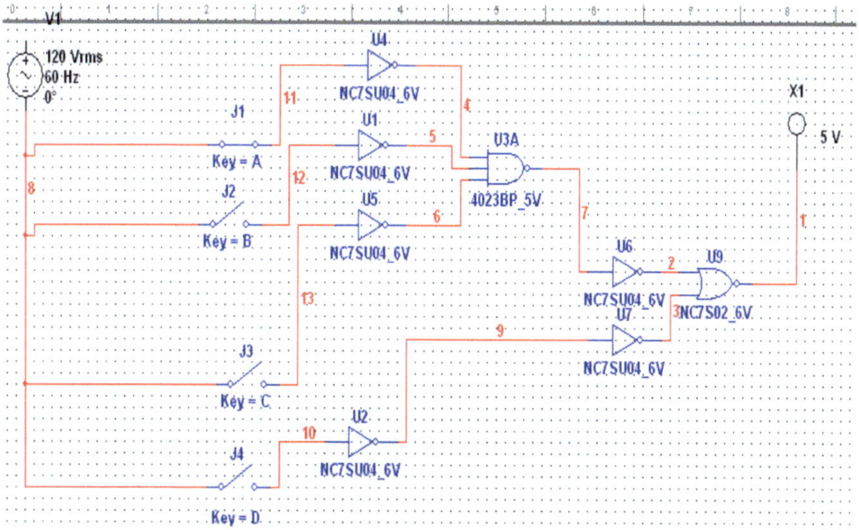

b) $Z=\overline{\overline{(\overline{\overline{\overline{A}\cdot\overline{B}\cdot\overline{C}}})+(\overline{\overline{D}})}}$

Fig. 7.17 The networks collected in Multisim 10

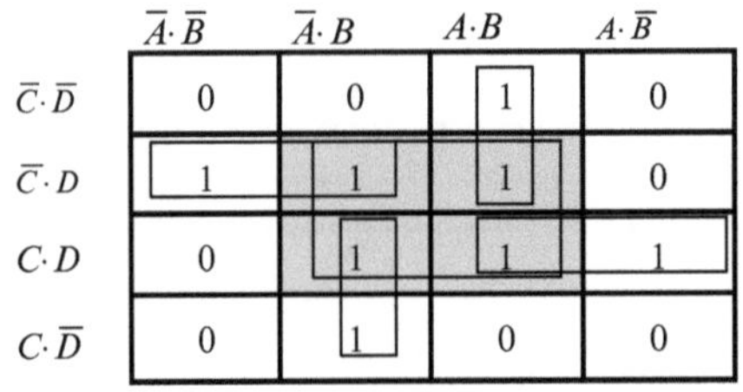

$$y = A\cdot B\cdot\overline{C} + \overline{A}\cdot B\cdot C + A\cdot C\cdot D + \overline{A}\cdot\overline{C}\cdot D.$$

Fig. 7.18 The Karnaugh Map for 4 variables

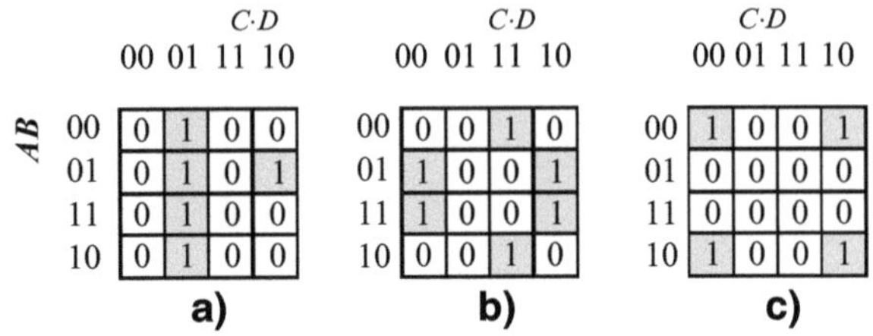

Fig. 7.19 Examples of minimization using Karnaugh maps for 4 variables

of four essential prime implicants and, therefore, must not be contained in any irredundant expression for y. We can thus summarize the procedure for obtaining a minimal sum-of-products expression for a function y.

1. Determine all essential prime implicants and include them in the minimal expression.
2. Remove from the list of prime implicants all those that are covered by the essential prime implicants.
3. If the set derived in step 1 covers all the minterms of f then it is the unique minimal expression. Otherwise, select additional prime implicants such that f is covered completely and such that the total number and size of the prime implicants thus added are minimal.

The execution of step 3 is not always straightforward. While in most cases with only a small number of variables this execution can be done by inspecting the map, in more complicated cases, and when the number of variables is large, a more systematic method is needed. The prime implicant chart presented in the next section is a possible tool aiding the search for a minimal expression (Fig. 7.19).

$$\text{Result}\quad (a) - y = f(A,B,C,D) = \overline{C}\cdot D \vee \overline{A}\cdot B\cdot C\cdot\overline{D}.$$
$$(b) - y = B\cdot\overline{D} \vee \overline{B}\cdot C\cdot D.$$
$$(c) - y = \overline{B}\cdot\overline{D}$$

7.3 On Quine-Mccluskey Method

The Quine-McCluskey, or Tabular, method is an algorithmic method that finds prime implicants, necessary prime implicants, and minimum sum-of-products expressions for digital systems with any number of variables.

The Quine-McCluskey algorithm was developed by W.V. Quine and Edward J. McCluskey. It is functionally identical to Karnaugh mapping, but the tabular form makes it more efficient for use in computer algorithms, and it also gives a deterministic way to check that the minimal form of a Boolean function has been reached.

The method involves two steps:

Finding all prime implicants of the function.

Use those prime implicants in a prime implicant chart to find the essential prime implicants of the function, as well as other prime implicants that are necessary to cover the function.

Steps in the Tabular method

1. Represent the minterms and the "don't cares" by the values of the input variables; For example, in a 5-variable system minterm 7 would be represented by *00111*;
2. Arrange the minterms and "don't cares" in groups according to the number of 1's:
   - Group with no 1's;
   - Group with a single 1;
   - Group with two 1's;
3. Compare each member of a group to each member of the adjacent group:
   - If two terms differ in a single position, then record the reduced expression and mark the two items as having been used.

   For example (Table 7.8)

   Or (Table 7.9)

   This step will identify all groups of two and all minterms/"don't cares" that are not members of a group of 2.

   Continue this process.
4. As you compare minterms and "don't cares", keep the reduced expressions in groups. For example

   Group from combining no "1"s with a single "1".
   Group from combining a single "1" with two "1"s.
   Group from combining two "1"s with three "1"s.
   Combine groups of two

Table 7.8 Truth table

A	B	C	$F(A,B,C)$
0	0	1	1
1	0	1	1

Table 7.9 The truth table after simplification

A	B	C	$F(A,B,C)$
*	0	1	1

Table 7.10 Truth table

№	A	B	C	$F(A,B,C)$
0	0	0	0	0
1	0	0	1	1
2	0	1	0	0
3	0	1	1	0
4	1	0	0	0
5	1	0	1	1
6	1	1	0	1
7	1	1	1	1

=>

Table 7.11 The truth table after simplification

№	A	B	C	$F(A,B,C)$
1	0	0	1	1
5	1	0	1	1
6	1	1	0	1
7	1	1	1	1

=>

Table 7.12 The truth table after simplification (minterm)

№	A	B	C	$F(A,B,C)$	**Minterm**
1	*	0	1	1	$\overline{B}{\cdot}C$
6	1	1	*	1	$A{\cdot}B$

5. Check each group of two expressions against each expression in the adjacent group.

Comparison is easier at this step and subsequent steps, because simplification is only possible if the groups of two have eliminated the same variable.

For example (Tables 7.10, 7.11, and 7.12)

The final expression is

$$F(A,B,C) = \overline{B}{\cdot}C + A{\cdot}B.$$

Table 7.13 Truth table

№	*A*	*B*	*C*	*D*	*S(A,B,C,D)*	**Minterm**	Number of 1's
0	0	0	0	0	0		
1	0	0	0	1	0		
2	0	0	1	0	0		
3	0	0	1	1	0		
4	0	1	0	0	1	*A′BC′D′*	1
5	0	1	0	1	1	*A′BC′D*	2
6	0	1	1	0	0		
7	0	1	1	1	1	*A′BCD*	3
8	1	0	0	0	1	*AB′C′D′*	1
9	1	0	0	1	0		
10	1	0	1	0	1	*AB′CD′*	2
11	1	0	1	1	0		
12	1	1	0	0	1	*ABC′D′*	2
13	1	1	0	1	1	*ABC′D*	3
14	1	1	1	0	0		
15	1	1	1	1	1	*ABCD*	4

=>

Table 7.14 The truth table after simplification

№	Line number	*A*	*B*	*C*	*D*	Number of 1's
1	4	0	1	0	0	1
	8	1	0	0	0	1
2	5	0	1	0	1	2
	10	1	0	1	0	2
	12	1	1	0	0	2
3	7	0	1	1	1	3
	13	1	1	0	1	3
4	15	1	1	1	1	4

=>

Table 7.15 The truth table after simplification

Line number	*A*	*B*	*C*	*D*	Number of 1's
4,5	0	1	0	*	1
4,12	*	1	0	0	1
8,10	1	0	*	0	1
8,12	1	*	0	0	1
5,7	0	1	*	1	2
5,13	*	1	0	1	2
12,13	1	1	0	*	2
7,15	*	1	1	1	3
13,15	1	1	*	1	3

=>

Table 7.16 The truth table after simplification

Line number	*A*	*B*	*C*	*D*	Number of 1's
4,5,12,13	*	1	0	*	1
4,12,5,13	*	1	0	*	1
8,10	1	0	*	0	1
8,12	1	*	0	0	1
5,7,13,15	*	1	*	1	2
5, 13,7, 15	*	1	*	1	2

=>

Table 7.17 The truth table after simplification

Line number	*A*	*B*	*C*	*D*	Number of 1's	**Minterm**
4,5,12,13	*	1	0	*	1	$B'C$
8,10	1	0	*	0	1	$AB'D'$
8,12	1	*	0	0	1	$AC'D'$
5,7,13,15	*	1	*	1	2	BD

=>

Table 7.18 The truth table after simplification

Line number	*A*	*B*	*C*	*D*	Number of 1's	**Minterm**
4,5,12,13	*	1	0	*	1	$B'C$
8,10	1	0	*	0	1	$AB'D'$
5,7,13,15	*	1	*	1	2	BD

Continue comparing group to adjacent group and separating the simpler expressions that result until no more simplifications can be achieved. This is the set of prime implicants.

Make a chart of which minterms are in which prime implicants:

- The prime implicants define the rows;
- The minterms define the columns;
- The "don't cares" are excluded at this point.

Example (Tables 7.13, 7.14, 7.15, 7.16, 7.17, and 7.18)

$$S(A,B,C,D) = BC' + AB'D' + BD.$$

Although more practical than Karnaugh mapping when dealing with more than four variables, the Quine–McCluskey algorithm also has a limited range of use since the problem it solves is NP-hard: the runtime of the Quine-McCluskey algorithm grows exponentially with the number of variables. It can be shown that for a function of n variables the upper bound on the number of prime implicants is $3^n/n$. Functions with a large number of variables have to be minimized with potentially non-optimal heuristic methods, of which the Espresso heuristic logic minimizer is the de-facto standard.

Table 7.19 The prime implicant table

Group number	Kontermy 0-th order	Kontermy 1-th order	Kontermy 2-th order
0	–		
1	0100	01*0	***1*0**
		*100	
2	0011	0*11	
	0110	*011	
	1010	011*	
	1100	*110	
		101*	
		1*10	
		11*0	
3	0111		
	1011		
	1110		
4	–		

Example

Step 1 finding prime implicants.

Minimizing an arbitrary function:

$$Y = \overline{D}\cdot\overset{1}{\overline{C}}\cdot B\cdot A + \overline{D}\cdot\overset{2}{C}\cdot\overline{B}\cdot\overline{A} + \overline{D}\cdot\overset{3}{C}\cdot B\cdot\overline{A} + \overline{D}\cdot\overset{4}{C}\cdot B\cdot A$$
$$+ D\cdot\overset{5}{\overline{C}}\cdot B\cdot A + D\cdot\overset{6}{\overline{C}}\cdot B\cdot A + D\cdot\overset{7}{C}\cdot\overline{B}\cdot\overline{A} + D\cdot\overset{8}{C}\cdot B\cdot\overline{A}.$$

Or

$$Y = 0\overset{1}{0}11 + 0\overset{2}{1}00 + 0\overset{3}{1}10 + 0\overset{4}{1}11 + 1\overset{5}{0}10 + 1\overset{6}{0}11 + 1\overset{7}{1}00 + 1\overset{8}{1}10.$$

At this point, one can start combining minterms with other minterms. If two terms vary by only a single digit changing, that digit can be replaced with a * indicating that the digit doesn't matter (Table 7.19).

For the first and second groups:

$$\mathit{0100}\ u \quad \mathit{0110} \to 01{*}0;$$
$$\mathit{0100}\ u \quad \mathit{1100} \to {*}100;$$

For the second and third groups:

$$\mathit{0011}\ u \quad \mathit{0111} \to 0{*}11;$$
$$\mathit{0011}\ u \quad \mathit{1011} \to {*}011;$$
$$\mathit{0110}\ u \quad \mathit{0111} \to 011{*};$$
$$\mathit{0110}\ u \quad \mathit{1110} \to {*}110;$$
$$\mathit{1010}\ u \quad \mathit{1011} \to 101{*};$$
$$\mathit{1010}\ u \quad \mathit{1110} \to 1{*}10;$$
$$\mathit{1100}\ u \quad \mathit{1110} \to 11{*}0;$$

Table 7.20 The prime implicant table after simplification

Group number	Kontermy 0-th order	Kontermy 1-th order	Kontermy 2-th order
0	–		
1	0100	01*0	***1*0**
		*100	
2	0011	0*11	
	0110	*011	
	1010	011*	
	1100	*110	
		101*	
		1*10	
		11*0	
3	0111		
	1011		
	1110		
4	–		

Perform the bonding kontermy first and second order (Table 7.20)

$$01{*}0\,u \quad 11{*}0 \rightarrow {*}1{*}0;$$
$$*100\,u \quad {*}110 \rightarrow {*}1{*}0;$$

$$Y = {*}1{*}0 + 0{*}11 + {*}011 + 011{*} + 101{*} + 1{*}10.$$

Step 2: Prime implicant chart.

None of the terms can be combined any further than this, so at this point we construct an essential prime implicant table. Along the side goes the prime implicants that have just been generated, and along the top go the minterms specified earlier. The "don't care" terms are not placed on top – they are omitted from this section because they are not necessary inputs (Table 7.21).

Here, each of the essential prime implicants has been starred – the prime implicant *1*0 ($C{\cdot}\overline{A}$) can be 'covered' by the 0100 ($\overline{D}{\cdot}C{\cdot}\overline{B}{\cdot}\overline{A}$), 0110 ($\overline{D}{\cdot}C{\cdot}B{\cdot}\overline{A}$), 1100 ($D{\cdot}C{\cdot}\overline{B}{\cdot}\overline{A}$), 1110 ($D{\cdot}C{\cdot}B{\cdot}\overline{A}$). If a prime implicant is essential then, as would be expected, it is necessary to include it in the minimized Boolean equation. In some cases, the essential prime implicants do not cover all minterms, in which case additional procedures for chart reduction can be employed.

Drawing logical product of logical sums of individual columns implicants

$$(B1 + C1){\cdot}A1{\cdot}(A1 + D1){\cdot}(B1 + D1){\cdot}(E1 + F1){\cdot}(C1 + E1){\cdot}A1{\cdot}(A1 + F1)$$

Or

$$A1{\cdot}(B1 + C1{\cdot}D1){\cdot}(E1 + C1{\cdot}F1) =$$
$$= A1{\cdot}B1{\cdot}E1 + A1{\cdot}B1{\cdot}C1{\cdot}F1 + A1{\cdot}C1{\cdot}D1{\cdot}E1 + A1{\cdot}C1{\cdot}D1{\cdot}F1.$$

Table 7.21 Prime implicant chart

Prime implicants	1 0011	2 0100	3 0110	4 0111	5 1010	6 1011	7 1100	8 1110
$A1 = {*}1{*}0$		*	*				*	*
$B1 = 0{*}11$	*			*				
$C1 = {*}011$	*					*		
$D1 = 011{*}$			*	*				
$E1 = 101{*}$					*	*		
$F1 = 1{*}10$					*			*

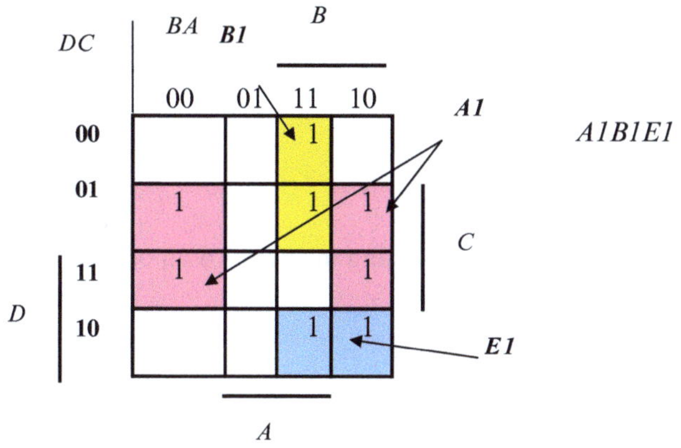

Fig. 7.20 The Veitch diagram

Instead of A1, B1, E1, we substitute the corresponding implicant –

$$y = {*}1{*}0 + 0{*}11 + 101{*}$$

The final expression is

$$Y = C{\cdot}\overline{A} + \overline{D}{\cdot}B{\cdot}A + D{\cdot}\overline{C}{\cdot}B.$$

Figures 7.20 and 7.21 show an example of a Karnaugh Map for both 4-variable expressions –

$$A1{\cdot}B1{\cdot}E1 + A1{\cdot}B1{\cdot}C1{\cdot}F1 + A1{\cdot}C1{\cdot}D1{\cdot}E1 + A1{\cdot}C1{\cdot}D1{\cdot}F1.$$

Example

Minimizing an arbitrary function

$$\overline{Y} = (D + C + \overset{1}{\overline{B}} + \overline{A}){\cdot}(D + \overset{2}{\overline{C}} + B + A){\cdot}(D + \overline{C} + \overset{3}{\overline{B}} + A){\cdot}(D + \overline{C} + \overset{4}{\overline{B}} + \overline{A}).$$

$${\cdot}(\overline{D} + \overset{5}{C} + \overline{B} + A){\cdot}(\overline{D} + \overset{6}{C} + \overline{B} + \overline{A}){\cdot}(\overline{D} + \overset{7}{\overline{C}} + B + A){\cdot}(\overline{D} + \overset{8}{\overline{C}} + \overline{B} + A).$$

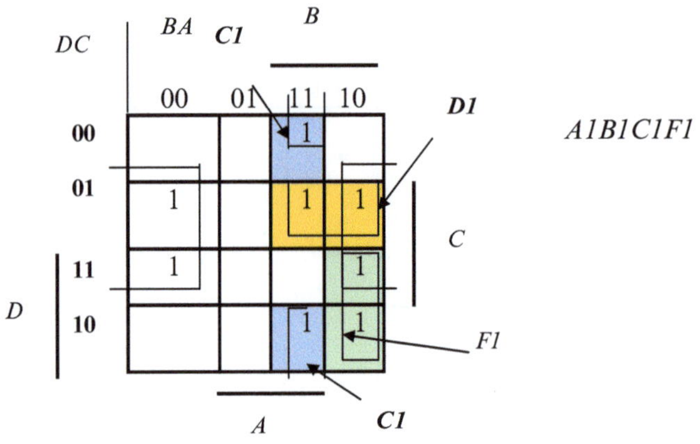

Fig. 7.21 The Veitch diagram

At this point, one can start combining maxterms with other maxterms. (Use the expression $(A+X)(A\cdot\overline{X})=A$).

1 and 4 maxterms =	$(D+C+\overline{B}+\overline{A})\cdot(D+\overline{C}^{4}+\overline{B}+\overline{A})\ =D+\overline{B}+\overline{A}.$	(1)
1 and 6 maxterms =	$(D+C+\overline{B}+\overline{A})\cdot(\overline{D}+C^{6}+\overline{B}+\overline{A})\ =C+\overline{B}+\overline{A}.$	(2)
2 and 3 maxterms =	$(D+\overline{C}+B+A)\cdot(D+\overline{C}^{3}+\overline{B}+A)\ =D+\overline{C}+A.$	(3)
2 and 7 maxterms =	$(D+\overline{C}+B+A)\cdot(\overline{D}+\overset{7}{\overline{C}}+B+A)=\overline{C}+B+A.$	(4)
3 and 4 maxterms =	$(D+\overline{C}+\overline{B}+A)\cdot(D+\overset{4}{\overline{C}}+\overline{B}+\overline{A})=D+\overline{C}+\overline{B}.$	(5)
3 and 8 maxterms =	$(D+\overline{C}+\overline{B}+A)\cdot(\overline{D}+\overline{C}^{8}+\overline{B}+A)\ =\overline{C}+\overline{B}+A.$	(6)
5 and 6 maxterms =	$(\overline{D}+C+\overline{B}+A)\cdot(\overline{D}+C^{6}+\overline{B}+\overline{A})\ =\overline{D}+C+\overline{B}.$	(7)
5 and 8 maxterms =	$(\overline{D}+C+\overline{B}+A)\cdot(\overline{D}+\overline{C}^{8}+\overline{B}+A)\ =\overline{D}+\overline{B}+A.$	(8)
7 and 8 maxterms =	$(\overline{D}+\overline{C}+B+A)\cdot(\overline{D}+\overline{C}^{8}+\overline{B}+A)\ =\overline{D}+\overline{C}+A.$	(9)

Using the expression $A\cdot A=A$, we obtain

$$\overline{Y}=(D+C+\overset{1}{\overline{B}}+\overline{A})\cdot(D+\overset{2}{\overline{C}}+B+A)\cdot(D+\overline{C}+\overset{3}{\overline{B}}+A)\cdot(D+\overline{C}+\overset{4}{\overline{B}}+\overline{A})\cdot$$

$$\cdot(\overline{D}+\overset{5}{C}+\overline{B}+A)\cdot(\overline{D}+\overset{6}{C}+\overline{B}+\overline{A})\cdot(\overline{D}+\overset{7}{\overline{C}}+B+A)\cdot(\overline{D}+\overset{8}{\overline{C}}+\overline{B}+A)\cdot$$

$$\cdot(D+\overset{1}{\overline{B}}+\overline{A})\cdot(C+\overset{2}{\overline{B}}+\overline{A})\cdot(D+\overset{3}{\overline{C}}+A)\cdot(\overline{C}+\overset{4}{B}+A)\cdot(D+\overset{5}{\overline{C}}+\overline{B})\cdot$$

$$\cdot(\overline{C}+\overset{6}{\overline{B}}+A)\cdot(\overline{D}+\overset{7}{C}+\overline{B})\cdot(\overline{D}+\overset{8}{\overline{B}}+A)\cdot(\overline{D}+\overset{9}{\overline{C}}+A)$$

Table 7.22 Prime implicant chart

	Maxterms							
Prime implicants	$D+C+\overline{B}+\overline{A}$	$D+\overline{C}+B+A$	$D+\overline{C}+\overline{B}+A$	$D+\overline{C}+\overline{B}+\overline{A}$	$\overline{D}+C+\overline{B}+A$	$\overline{D}+C+\overline{B}+\overline{A}$	$\overline{D}+\overline{C}+B+A$	$\overline{D}+\overline{C}+\overline{B}+A$
$\overline{C}+A$		*	*				*	*
$D+\overline{B}+\overline{A}$	*			*				
$C+\overline{B}+\overline{A}$	*					*		
$D+\overline{C}+\overline{B}$			*	*				
$\overline{D}+C+\overline{B}$					*	*		
$\overline{D}+\overline{B}+A$					*			*

Or, using the expression $A\cdot(A+B)=A$, we obtain

$$\overline{Y}=(D+\overset{1}{\overline{B}}+\overline{A})\cdot(C+\overset{2}{\overline{B}}+\overline{A})\cdot(D+\overset{3}{\overline{C}}+A)\cdot(\overline{C}+{}^{4}B+A)\cdot(D+\overset{5}{\overline{C}}+\overline{B})\cdot(\overline{C}+\overset{6}{\overline{B}}+A)\cdot$$
$$\cdot(\overline{D}+{}^{7}C+\overline{B})\cdot(\overline{D}+\overset{8}{\overline{B}}+A)\cdot(\overline{D}+\overset{9}{\overline{C}}+A)$$

Using the expression $(A+X)\cdot(A+\overline{X})=A$, we obtain

3 and 9 maxterms = $(D+\overset{3}{\overline{C}}+A)\cdot(\overline{D}+\overset{9}{\overline{C}}+A)=\overline{C}+A.$

4 and 6 maxterms = $(\overline{C}+\overset{4}{B}+A)\cdot(\overline{C}+\overset{6}{\overline{B}}+A)=\overline{C}+A.$

Using the expression $A\cdot A=A$, we obtain

$$\overline{Y}=(D+\overset{1}{\overline{B}}+\overline{A})\cdot(C+\overset{2}{\overline{B}}+\overline{A})\cdot(D+\overset{3}{\overline{C}}+A)\cdot(\overline{C}+{}^{4}B+A)\cdot(D+\overset{5}{\overline{C}}+\overline{B})\cdot(\overline{C}+\overset{6}{\overline{B}}+A)\cdot$$
$$\cdot(\overline{D}+{}^{7}C+\overline{B})\cdot(\overline{D}+\overset{8}{\overline{B}}+A)\cdot(\overline{D}+\overset{9}{\overline{C}}+A)\cdot(\overline{C}+A).$$

Using the expression $A\cdot(A+B)=A$, we obtain

$$\overline{Y}=(D+\overset{1}{\overline{B}}+\overline{A})\cdot(C+\overset{2}{\overline{B}}+\overline{A})\cdot(D+\overset{5}{\overline{C}}+\overline{B})\cdot(\overline{D}+{}^{7}C+\overline{B})\cdot(\overline{D}+\overset{8}{\overline{B}}+A)\cdot(\overline{C}+A).$$

Apply Step 2: prime implicant chart (Table 7.22)

Here, each of the essential prime implicants has been starred – the prime implicant $\overline{C}+A$ can be 'covered' by the maxterms $D+\overline{C}+B+A$, $D+\overline{C}+\overline{B}+A$, $\overline{D}+\overline{C}+B+A$, $\overline{D}+\overline{C}+\overline{B}+A$.

The final implicant chart (Table 7.23)

The final expression is

$$\overline{Y}=(\overline{C}+A)\cdot(D+\overline{B}+\overline{A})\cdot(\overline{D}+C+\overline{B}).$$

Table 7.23 The final implicant chart

Prime implicants	Maxterms							
	$D+C+$ $+\overline{B}+\overline{A}$	$D+\overline{C}+$ $+B+A$	$D+\overline{C}+$ $+\overline{B}+A$	$D+\overline{C}+$ $+\overline{B}+\overline{A}$	$\overline{D}+C+$ $+\overline{B}+A$	$\overline{D}+C+$ $+\overline{B}+\overline{A}$	$\overline{D}+\overline{C}+$ $+B+A$	$\overline{D}+\overline{C}+$ $+\overline{B}+A$
$\overline{C}+A$		*	*				*	*
$D+\overline{B}+\overline{A}$	*			*				
$\overline{D}+C+\overline{B}$					*	*		

Or

$$\overline{Y} = (\overline{C}+A)\cdot(D+\overline{B}+\overline{A})\cdot(\overline{D}+C+\overline{B})$$

$$Y = \overline{\overline{C}+A} + \overline{D+\overline{B}+\overline{A}} + \overline{\overline{D}+C+\overline{B}} = C\cdot\overline{A} + \overline{D}\cdot B\cdot A + D\cdot\overline{C}\cdot B.$$

Example
Minimizing an arbitrary function

$$Y = \overline{D}\cdot\overline{C}^{1}\cdot B\cdot A + \overline{D}\cdot C^{2}\cdot\overline{B}\cdot\overline{A} + \overline{D}\cdot C^{3}\cdot B\cdot\overline{A} + \overline{D}\cdot C^{4}\cdot B\cdot A + D\cdot\overline{C}^{5}\cdot B\cdot\overline{A} + D\cdot\overline{C}^{6}\cdot B\cdot A + \\ + D\cdot C^{7}\cdot\overline{B}\cdot\overline{A} + D\cdot C^{8}\cdot B\cdot\overline{A}.$$

At this point, one can start combining minterms with other minterms. (Use the expression $A\cdot X + A\cdot\overline{X} = A$).

1 and 4 minterms =	$\overline{D}\cdot\overline{C}^{1}\cdot B\cdot A + \overline{D}\cdot C^{4}\cdot B\cdot A = \overline{D}\cdot B\cdot A.$	(1)
1 and 6 minterms =	$\overline{D}\cdot\overline{C}^{1}\cdot B\cdot A + D\cdot\overline{C}^{6}\cdot B\cdot A = \overline{C}\cdot B\cdot A.$	(2)
2 and 3 minterms =	$\overline{D}\cdot C^{2}\cdot\overline{B}\cdot\overline{A} + \overline{D}\cdot C^{3}\cdot B\cdot\overline{A} = \overline{D}\cdot C\cdot\overline{A}.$	(3)
2 and 7 minterms =	$\overline{D}\cdot C^{2}\cdot\overline{B}\cdot\overline{A} + D\cdot C^{7}\cdot\overline{B}\cdot\overline{A} = C\cdot\overline{B}\cdot\overline{A}.$	(4)
3 and 4 minterms =	$\overline{D}\cdot C^{3}\cdot B\cdot\overline{A} + \overline{D}\cdot C^{4}\cdot B\cdot A = \overline{D}\cdot C\cdot B.$	(5)
3 and 8 minterms =	$\overline{D}\cdot C^{3}\cdot B\cdot\overline{A} + D\cdot C^{8}\cdot B\cdot\overline{A} = C\cdot B\cdot\overline{A}.$	(6)
5 and 6 minterms =	$D\cdot\overline{C}^{5}\cdot B\cdot\overline{A} + D\cdot\overline{C}^{6}\cdot B\cdot A = D\cdot\overline{C}\cdot B.$	(7)
5 and 8 minterms =	$D\cdot\overline{C}^{5}\cdot B\cdot\overline{A} + D\cdot C^{8}\cdot B\cdot\overline{A} = D\cdot B\cdot\overline{A}.$	(8)
7 and 8 minterms =	$D\cdot C^{7}\cdot\overline{B}\cdot\overline{A} + D\cdot C^{8}\cdot B\cdot\overline{A} = D\cdot C\cdot\overline{A}.$	(9)

Using the expression $A + A = A$, we obtain

$$Y = \overline{D}\cdot\overline{C}^{1}\cdot B\cdot A + \overline{D}\cdot C^{2}\cdot\overline{B}\cdot\overline{A} + \overline{D}\cdot C^{3}\cdot B\cdot\overline{A} + \overline{D}\cdot C^{4}\cdot B\cdot A + D\cdot\overline{C}^{5}\cdot B\cdot\overline{A} + D\cdot\overline{C}^{6}\cdot B\cdot A + \\ + D\cdot C^{7}\cdot\overline{B}\cdot\overline{A} + D\cdot C^{8}\cdot B\cdot\overline{A} + \overline{D}\cdot B\cdot A + \overline{C}\cdot B\cdot A + \overline{D}\cdot C\cdot\overline{A} + C\cdot\overline{B}\cdot\overline{A} + \overline{D}\cdot C\cdot B + \\ + C\cdot B\cdot\overline{A} + D\cdot\overline{C}\cdot B + D\cdot B\cdot\overline{A} + D\cdot C\cdot\overline{A}.$$

Table 7.24 Prime implicant chart

Prime implicants	Minterms							
	$\overline{D}\cdot\overline{C}\cdot B\cdot A$	$\overline{D}\cdot C\cdot\overline{B}\cdot\overline{A}$	$\overline{D}\cdot C\cdot B\cdot\overline{A}$	$\overline{D}\cdot C\cdot B\cdot A$	$D\cdot\overline{C}\cdot B\cdot\overline{A}$	$D\cdot\overline{C}\cdot B\cdot A$	$D\cdot C\cdot\overline{B}\cdot\overline{A}$	$D\cdot C\cdot B\cdot\overline{A}$
$C\cdot\overline{A}$		*	*				*	*
$\overline{D}\cdot B\cdot A$	*			*				
$\overline{C}\cdot B\cdot A$	*					*		
$\overline{D}\cdot C\cdot B$			*	*				
$D\cdot\overline{C}\cdot B$					*	*		
$D\cdot B\cdot\overline{A}$					*			*

Table 7.25 The final implicant chart

Prime implicants	Minterms							
	$\overline{D}\cdot\overline{C}\cdot B\cdot A$	$\overline{D}\cdot C\cdot\overline{B}\cdot\overline{A}$	$\overline{D}\cdot C\cdot B\cdot\overline{A}$	$\overline{D}\cdot C\cdot B\cdot A$	$D\cdot\overline{C}\cdot B\cdot\overline{A}$	$D\cdot\overline{C}\cdot B\cdot A$	$D\cdot C\cdot\overline{B}\cdot\overline{A}$	$D\cdot C\cdot B\cdot\overline{A}$
$C\cdot\overline{A}$		*	*				*	*
$\overline{D}\cdot B\cdot A$	*			*				
$D\cdot\overline{C}\cdot B$					*	*		

Using the expression $A + A\cdot B = A$, we obtain

$$Y = \overline{D}\cdot{}^{1}B\cdot A + \overline{C}\cdot{}^{2}B\cdot A + \overline{D}\cdot{}^{3}C\cdot\overline{A} + C\cdot\overset{4}{\overline{B}}\cdot\overline{A} + \overline{D}\cdot{}^{5}C\cdot B +$$
$$+C\cdot{}^{6}B\cdot\overline{A} + D\cdot\overset{7}{\overline{C}}\cdot B + D\cdot{}^{8}B\cdot\overline{A} + D\cdot{}^{9}C\cdot\overline{A}.$$

Using the expression $A\cdot X + A\cdot\overline{X} = A$, we obtain
3 and 9 minterms = $\overline{D}\cdot\underset{3}{{}^{3}C}\cdot\overline{A} + D\cdot\underset{6}{{}^{9}C}\cdot\overline{A} = C\cdot\overline{A}$.
4 and 6 minterms = $C\cdot\overline{B}\cdot\overline{A} + B\cdot C\cdot\overline{A} = C\cdot\overline{A}$.
Using the expression $A + A = A$, we obtain

$$Y = \overline{D}\cdot{}^{1}B\cdot A + \overline{C}\cdot{}^{2}B\cdot A + \overline{D}\cdot{}^{3}C\cdot\overline{A} + C\cdot\overset{4}{\overline{B}}\cdot\overline{A} + \overline{D}\cdot{}^{5}C\cdot B +$$
$$+C\cdot{}^{6}B\cdot\overline{A} + D\cdot\overset{7}{\overline{C}}\cdot B + D\cdot\overset{8}{B}\cdot\overline{A} + D\cdot{}^{9}C\cdot\overline{A} + C\cdot\overline{A}.$$

Using the expression $A + A\cdot B = A$, we obtain

$$Y = \overline{D}\cdot{}^{1}B\cdot A + \overline{C}\cdot{}^{2}B\cdot A + \overline{D}\cdot{}^{5}C\cdot B + D\cdot\overset{7}{\overline{C}}\cdot B + D\cdot{}^{8}B\cdot\overline{A} + C\cdot\overline{A}.$$

Apply Step 2: prime implicant chart (Table 7.24)
Here, each of the essential prime implicants has been starred – the prime implicant $C\cdot\overline{A}$ can be 'covered' by the minterms $\overline{D}\cdot C\cdot\overline{B}\cdot\overline{A}$, $\overline{D}\cdot C\cdot B\cdot\overline{A}$, $D\cdot C\cdot\overline{B}\cdot\overline{A}$, $D\cdot C\cdot B\cdot\overline{A}$.
The final implicant chart (Table 7.25)

Table 7.26 Minimization of Boolean expression

Variant	Boolean expression
1	$y = A{\cdot}(\overline{B}{\cdot}\overline{D} + D) + A{\cdot}\overline{B}{\cdot}C + \overline{D} + \overline{A{\cdot}(\overline{B}{\cdot}\overline{D} + C)}.$
2	$y = \overline{A}{\cdot}\overline{B}{\cdot}\overline{D} + A{\cdot}B{\cdot}D + \overline{A{\cdot}(\overline{B}{\cdot}\overline{D} + C)} + A{\cdot}\overline{B}{\cdot}C + \overline{D}.$
3	$y = \overline{A}{\cdot}\overline{B}{\cdot}\overline{D} + \overline{A{\cdot}(\overline{B}{\cdot}\overline{D} + C)} + A{\cdot}B{\cdot}D + \overline{D} + A{\cdot}\overline{B}{\cdot}\overline{C}.$
4	$y = A{\cdot}\overline{B}{\cdot}C + A{\cdot}(\overline{B}{\cdot}\overline{D} + C) + \overline{D}{\cdot}A + \overline{A{\cdot}(\overline{B}{\cdot}\overline{D} + \overline{C})}.$
5	$y = \overline{D}{\cdot}A + \overline{A{\cdot}(\overline{B}{\cdot}\overline{D} + \overline{C})} + A{\cdot}\overline{B} + A{\cdot}(\overline{B}{\cdot}\overline{D} + C) + \overline{D}{\cdot}A{\cdot}\overline{B}{\cdot}C.$
6	$y = \overline{D}{\cdot}A + \overline{A{\cdot}(\overline{B}{\cdot}\overline{D} + \overline{C})} + A{\cdot}\overline{B} + A{\cdot}(\overline{B}{\cdot}\overline{D} + C) + D{\cdot}\overline{A}{\cdot}\overline{B}{\cdot}\overline{C}.$
7	$y = \overline{D}{\cdot}A + \overline{A{\cdot}(\overline{B}{\cdot}\overline{D} + \overline{C})} + \overline{A}{\cdot}\overline{B} + A{\cdot}(\overline{B}{\cdot}\overline{D} + C) + \overline{D}{\cdot}A{\cdot}\overline{B}{\cdot}C + \overline{A}{\cdot}B.$
8	$y = \overline{A}{\cdot}\overline{B}{\cdot}D + A{\cdot}(\overline{B}{\cdot}\overline{D} + C) + \overline{D}{\cdot}A{\cdot}\overline{B}{\cdot}\overline{C} + \overline{A}{\cdot}B{\cdot}C + \overline{D}{\cdot}A + \overline{A{\cdot}(\overline{B}{\cdot}\overline{D} + \overline{C})}.$
9	$y = A{\cdot}(\overline{B}{\cdot}\overline{D} + C) + \overline{D}{\cdot}A{\cdot}\overline{B}{\cdot}\overline{C} + \overline{A}{\cdot}B{\cdot}\overline{C} + A{\cdot}\overline{D} + C.$
10	$y = \overline{D}{\cdot}A + \overline{A{\cdot}(\overline{B}{\cdot}\overline{D} + \overline{C})} + \overline{C}(\overline{A}{\cdot}\overline{B}{\cdot}\overline{D} + D) + A{\cdot}\overline{B}{\cdot}C + \overline{D}.$
11	$y = A{\cdot}(\overline{B}{\cdot}\overline{D} + C) + \overline{D}{\cdot}(A{\cdot}\overline{B}{\cdot}\overline{C} + \overline{A}{\cdot}B{\cdot}\overline{C}).$
12	$y = \overline{D}{\cdot}A + \overline{A{\cdot}(\overline{B} + \overline{C})} + A{\cdot}(\overline{B}{\cdot}\overline{D} + C) + D{\cdot}(A{\cdot}B{\cdot}\overline{C} + \overline{A}{\cdot}B{\cdot}\overline{C}).$
13	$y = \overline{D}{\cdot}A + \overline{A{\cdot}(\overline{B} + \overline{C})} + B{\cdot}(A{\cdot}\overline{D} + C) + \overline{D}{\cdot}(\overline{B}{\cdot}\overline{C}{\cdot}A + \overline{A}{\cdot}B{\cdot}C) + D.$
14	$y = A{\cdot}\overline{B}{\cdot}\overline{D}{\cdot}C + \overline{D}{\cdot}A + \overline{A{\cdot}(\overline{B} + \overline{C})} + A{\cdot}(\overline{B}{\cdot}\overline{D} + C{\cdot}D) + D{\cdot}\overline{A}{\cdot}\overline{B}{\cdot}\overline{C}.$
15	$y = C{\cdot}(\overline{A}{\cdot}\overline{D} + B) + \overline{D}{\cdot}(A{\cdot}\overline{B}{\cdot}\overline{C} + \overline{A}{\cdot}B{\cdot}\overline{C}).$
16	$y = A{\cdot}\overline{B}{\cdot}\overline{D} + C{\cdot}\overline{D} + A{\cdot}\overline{B}{\cdot}\overline{C} + \overline{A}{\cdot}B{\cdot}\overline{C} + \overline{D}{\cdot}A + \overline{A{\cdot}(\overline{B} + \overline{C})}.$
17	$y = (\overline{C + D}) + A{\cdot}\overline{C}{\cdot}\overline{D} + A{\cdot}\overline{B}{\cdot}\overline{C} + \overline{A}{\cdot}\overline{B}{\cdot}C{\cdot}D + \overline{D}{\cdot}A + \overline{A{\cdot}(\overline{B} + \overline{C})} + A{\cdot}C{\cdot}D.$
18	$y = A{\cdot}B{\cdot}(\overline{\overline{C}{\cdot}D}) + A{\cdot}\overline{B}{\cdot}D + \overline{B}{\cdot}\overline{C}{\cdot}\overline{D} + \overline{D}{\cdot}A + \overline{A{\cdot}(\overline{B} + \overline{C})}.$
19	$y = A{\cdot}\overline{B} + \mathrm{A}{\cdot}\overline{D} + A{\cdot}\overline{B}{\cdot}\overline{C} + \overline{A}{\cdot}B{\cdot}\overline{C} + \overline{D}{\cdot}A + \overline{A{\cdot}(\overline{B} + \overline{C})}.$
20	$y = A{\cdot}(\overline{B}{\cdot}\overline{D} + C) + \overline{D}{\cdot}A{\cdot}\overline{B}{\cdot}\overline{C} + \overline{A}{\cdot}B{\cdot}\overline{C} + A{\cdot}\overline{D} + C.$
21	$y = \overline{A{\cdot}(\overline{B}{\cdot}\overline{D} + \overline{C})} + \overline{A}{\cdot}\overline{B} + A{\cdot}(\overline{B}{\cdot}\overline{D} + D) + \overline{D}{\cdot}A{\cdot}\overline{B}{\cdot}C + \overline{A}{\cdot}B.$
22	$y = \overline{A}{\cdot}\overline{B}{\cdot}\overline{D} + A{\cdot}B{\cdot}D + \overline{A{\cdot}(\overline{B}{\cdot}\overline{D} + C)} + A{\cdot}\overline{B}{\cdot}C + \overline{D}.$
23	$y = D{\cdot}(\overline{B}{\cdot}\overline{D} + A) + \overline{D}{\cdot}(A{\cdot}\overline{B}{\cdot}\overline{C} + \overline{A}{\cdot}B{\cdot}\overline{C}).$
24	$y = A{\cdot}(\overline{B}{\cdot}\overline{D} + C{\cdot}\overline{D}) + A{\cdot}\overline{B}{\cdot}C + \overline{D} + \overline{A{\cdot}(\overline{B}{\cdot}\overline{D} + C)}.$
25	$y = A{\cdot}(\overline{B}{\cdot}\overline{D} + \overline{A}{\cdot}D) + A{\cdot}\overline{B}{\cdot}C + \overline{D} + \overline{A{\cdot}(\overline{B}{\cdot}\overline{D} + C)}.$

The final expression is $Y = C{\cdot}\overline{A} + \overline{D}{\cdot}B{\cdot}A + D{\cdot}\overline{C}{\cdot}B.$

Below are options for logical expressions. Perform the minimization of these functions using – the laws of Boolean algebra, the Karnaugh Map method and the implicant charts (Table 7.26).

Chapter 8
Latches, Flip-Flops, Counters, Registers, Timer, Multiplexer, Decoder, Etc.

Abstract The chapter analyzes the fundamentals of sequential logic, like categories of bistable devices (i.e. the latch and the flip-flop) and more complex circuits built from these basic blocks, like the basic RS NAND latch, the clocked RS NAND latch, a basic digital counter, a synchronous binary counter, BCD counter, the Johnson counter, serial-to-parallel shift register, parallel-to-serial shift register, timer, the multiplexer, the demultiplexer, digital comparator and the digital encoder and decoder.

8.1 Latches

This chapter begins a study of the fundamentals of sequential logic. We study two categories of bistable devices, the latch and the flip-flop. Bistable devices have two stable states, called SET and RESET; they can retain either of these states indefinitely, making them useful as storage devices. The basic difference between latches and flip-flops is the way in which they are changed from one state to the other. The flip-flop is a basic building block for counters, registers, and other sequential control logic and is used in certain types of memories. The monostable multivibrator, commonly known as the one-shot, has only one stable state. A one-shot produces a single controlled-width pulse when activated or triggered. The astable multivibrator has no stable state and is used primarily as an oscillator, which is a self-sustained waveform generator. Pulse oscillators are used as the sources for timing waveforms in digital systems.

In electronics, a flip-flop is a circuit that has two stable states and can be used to store state information.

A flip-flop is usually controlled by control signals that can include a clock signal. The outputs usually include the complement as well as the normal output.

B. Borowik et al., *Theory of Digital Automata*, Intelligent Systems, Control and Automation: Science and Engineering 63, DOI 10.1007/978-94-007-5228-3_8,

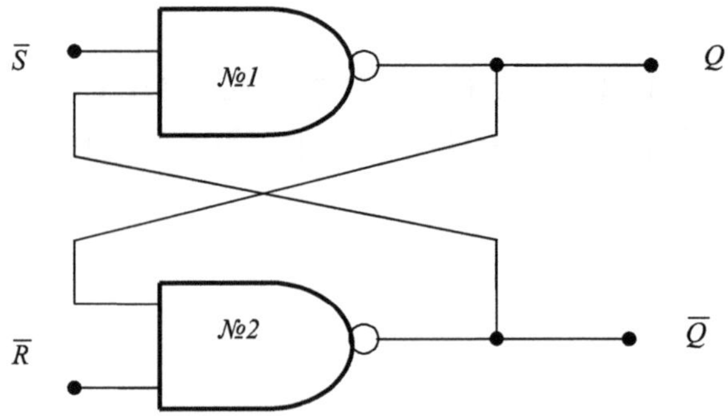

Fig. 8.1 RS NAND Latch

8.1.1 The Basic RS NAND Latch

In order for a logical circuit to "remember" and retain its logical state even after the controlling input signal(s) have been removed, it is necessary for the circuit to include some form of feedback. We might start with a pair of inverters, each having its input connected to the other's output. The two outputs will always have opposite logic levels.

The problem with this is that we don't have any additional inputs that we can use to change the logic states if we want. We can solve this problem by replacing the inverters with NAND or NOR gates, and using the extra input lines to control the circuit.

The circuit shown below is a basic NAND latch. The inputs are generally designated "**S**" and "**R**" for "**Set**" and "**Reset**" respectively. Because the NAND inputs must normally be logic 1 to avoid affecting the latching action, the inputs are considered to be inverted in this circuit.

For the NAND latch circuit, both inputs should normally be at a logic 1 level. Changing an input to a logic 0 level will force that output to a logic 1. The same logic 1 will also be applied to the second input of the other NAND gate, allowing that output to fall to a logic 0 level. This in turn feeds back to the second input of the original gate, forcing its output to remain at logic 1.

Applying another logic 0 input to the same gate will have no further effect on this circuit. However, applying a logic 0 to the other gate will cause the same reaction in the other direction, thus changing the state of the latch circuit the other way (Fig. 8.1).

Note that it is forbidden to have both inputs at a logic 0 level at the same time. That state will force both outputs to a logic 1, overriding the feedback latching action. In this condition, whichever input goes to logic 1 first will lose control, while the other input (still at logic 0) controls the resulting state of the latch. If both inputs go to logic 1 simultaneously, the result is a "race" condition, and the final state of the latch cannot be determined ahead of time.

An active-LOW input S-R latch is formed with two cross-coupled NAND gates, as shown in Fig. 8.2. Notice that the output of each gate is connected to an input of the opposite gate. This produces the regenerative feedback that is characteristic of all latches and flip-flops.

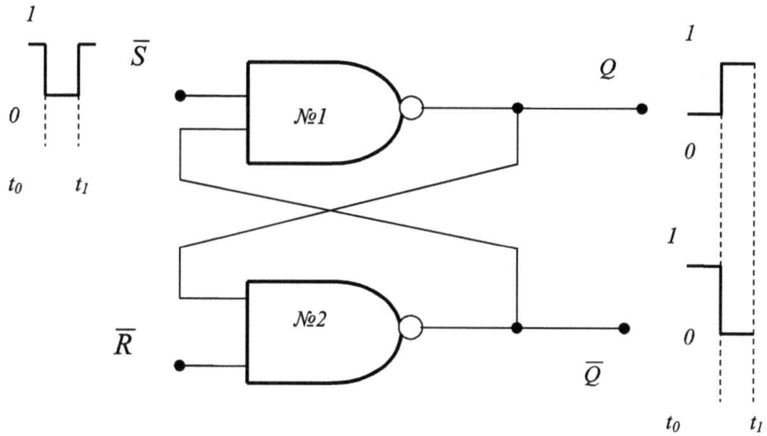

Fig. 8.2 An active-LOW input RS latch is formed with two cross-coupled NAND gates

Table 8.1 Gated RS NAND latch truth table

SET	*CLEAR*	*Quit*
1	1	No change
0	1	$Q = 1$
1	0	$Q = 0$
0	0	Not allowed ($Q = \overline{Q} = 1$).

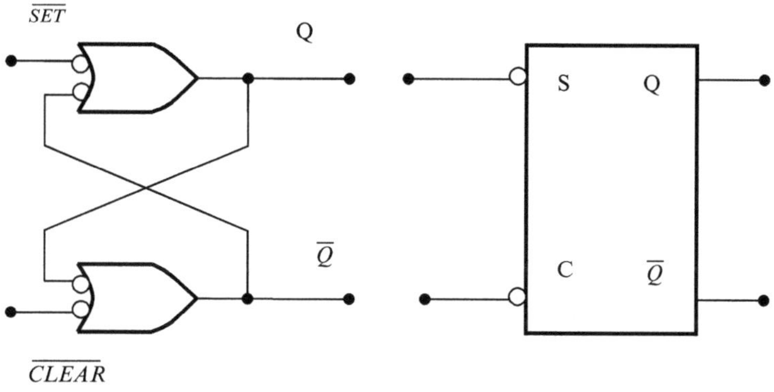

Fig. 8.3 Latch output remembers the last input that was activated and will not change states until the opposite input is activated

Truth table for the basic RS NAND latch (Table 8.1).

If the $\overline{S}$ and $\overline{R}$ waveforms in Fig. 8.3 are applied to the inputs of the latch in Fig. 8.4, determine the waveform that will be observed on the Q output. Assume that Q is initially LOW.

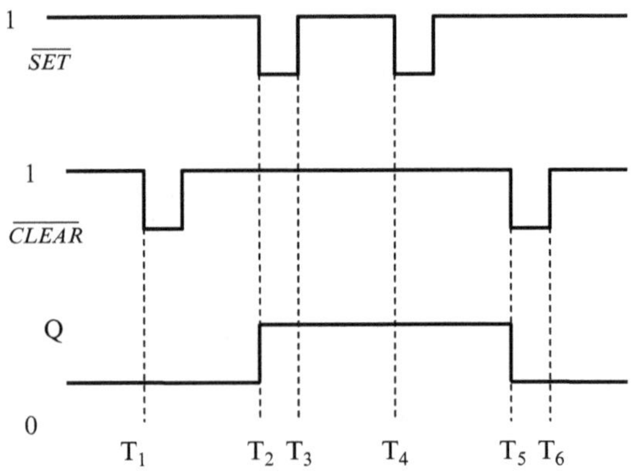

Fig. 8.4 The waveforms for example basic RS NAND latch

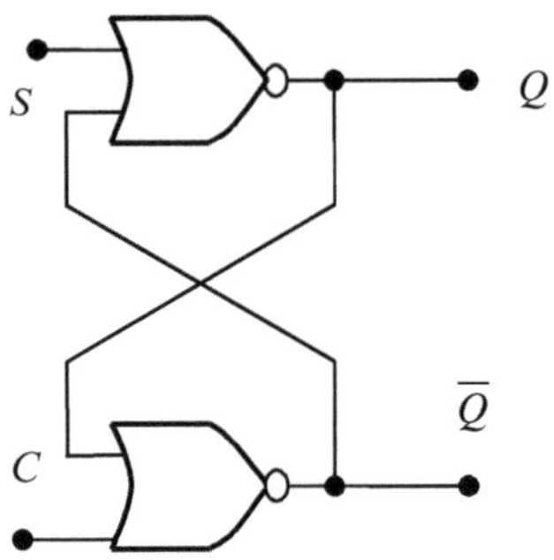

Fig. 8.5 RS flip-flop composed of two NOR gates

Table 8.2 Gated RS NOR latch truth table

SET	*CLEAR*	*Quit*
0	0	No change
1	0	$Q = 1$
0	1	$Q = 0$
1	1	Not allowed ($Q = \overline{Q} = 0$).

8.1.2 The Basic RS NOR Latch

While most of our demonstration circuits use NAND gates, the same functions can also be performed using NOR gates. A few adjustments must be made to allow for the difference in the logic function, but the logic involved is quite similar.

The circuit shown below is a basic NOR latch. The inputs are generally designated "**S**" and "**R**" for "**Set**" and "**Reset**" respectively. Because the NOR inputs must normally be logic 0 to avoid overriding the latching action, the inputs are not inverted in this circuit.

An active-HIGH input S-R (SET-RESET) latch is formed with two cross-coupled NOR gates, as shown in Fig. 8.5.

Truth table for the basic RS NOR latch (Table 8.2).

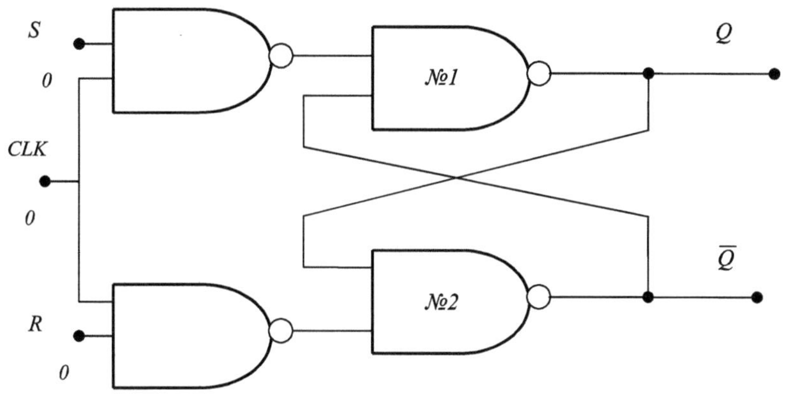

Fig. 8.6 The Clocked RS NAND latch

One problem with the basic **RS NOR** latch is that the input signals actively drive their respective outputs to a logic 0, rather than to a logic 1. Thus, the S input signal is applied to the gate that produces the $\overline{Q}$ output, while the R input signal is applied to the gate that produces the Q output. The circuit works fine, but this reversal of inputs can be confusing when you first try to deal with NOR-based circuits.

8.1.3 The Clocked RS NAND Latch

By adding a pair of NAND gates to the input circuits of the RS latch, we accomplish two goals: normal rather than inverted inputs and a third input common to both gates which we can use to synchronize this circuit with others of its kind, Fig. 8.6.

The clocked RS latch circuit is very similar in operation to the basic latch. The S and R inputs are normally at logic 0, and must be changed to logic 1 to change the state of the latch. However, with the third input, a new factor has been added. This input is typically designated C or CLK, because it is typically controlled by a clock circuit of some sort, which is used to synchronize several of these latch circuits with each other. The output can only change state while the CLK input is a logic 1. When CLK is a logic 0, the S and R inputs will have no effect.

The same rule about not activating both the S and R inputs simultaneously holds true: if both are logic 1 when the clock is also logic 1, the latching action is bypassed and both outputs will go to logic 1. The difference in this case is that if the CLK input drops to logic 0 first, there is no question or doubt – a true race condition will exist, and you cannot tell which way the outputs will come to rest. The example circuit on this page reflects this uncertainty.

For correct operation, the selected R or S input should be brought to logic 1, then the CLK input should be made logic 1 and then logic 0 again. Finally, the selected input should be returned to logic 0.

Table 8.3 Flip-flop symbols

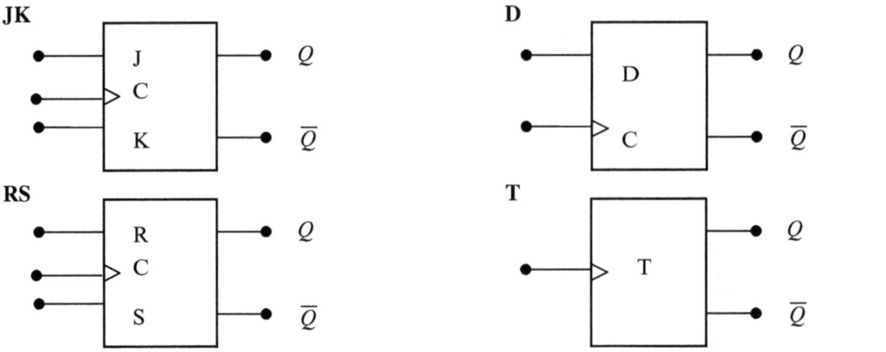

8.2 Edge-Triggered Flip-Flops

Although the internal circuitry of latches is interesting to watch on an individual basis, placing all of those logic symbols in a diagram involving multiple flip-flops would rapidly generate so much clutter that the overall purpose of the diagram would be lost. Now we are using one symbol to represent a cluster of logic gates connected to perform a specific function.

Flip-flops are synchronous bistable devices, also known as bistable multivibrators. In this case, the term synchronous means that the output changes state only at a specified point on the triggering input called the clock (CLK), which is designated as a control input, C; that is, changes in the output occur in synchronization with the clock.

8.2.1 Flip-Flop Symbols

An edge-triggered flip-flop changes state either at the positive edge (rising edge) or at the negative edge (falling edge) of the clock pulse and is sensitive to its inputs only at this transition of the clock. Three types of edge-triggered flip-flops are covered in this section: SR, D, and JK. Although the S-R flip-flop is not available in IC form, it is the basis for the D and J-K flip-flops. The logic symbols for all of these flip-flops are shown in Table 8.3.

As you have no doubt noticed, the symbols above are nearly identical – only the inputs vary. In each of the symbols above, the clock input is marked by the small angle, rather than by the letters CLK. That little angle marker actually provides two pieces of information, rather than one. First, of course, it marks the clocking input. Second, it specifies that these are edge-triggered flip-flops.

Any of these symbols may be modified according to their actual use within the larger circuit.

It is very seldom that a flip-flop will actually be used alone. Such circuits are far more useful when grouped together and acting in concert. There are two general ways in which flip-flops may be interconnected to perform useful functions: counters

Table 8.4 Truth table for a positive edge-triggered S-R flip-flop

INPUTS			OUTPUTS		
S	*R*	*CLK*	*Q*	$\bar{Q}$	COMMENTS
0	0	X	Q_0	$\bar{Q}_0$	No change
0	1	↑	0	1	RESET
1	0	↑	1	0	SET
1	1	↑	?	?	Invalid

↑ = clock transition LOW to HIGH
X = irrelevant ("don't care")
Q_0 = output level prior to clock transition

Table 8.5 Truth table for a positive edge-triggered S-C flip-flop

INPUTS			OUTPUT	
S	*C*	*CLK*	*Q*	COMMENTS
0	0	↑	Q_0	No change
1	0	↑	1	RESET
0	1	↑	0	SET
1	1	↑	?	

Table 8.6 Truth table for a positive edge-triggered J-K flip-flop

INPUTS			OUTPUTS		
J	*K*	*CLK*	*Q*	$\bar{Q}$	COMMENTS
0	0	↑	*Q*	$\bar{Q}_0$	No change
0	1	↑	0	1	RESET
1	0	↑	1	0	SET
1	1	↑	$\bar{Q}_0$	Q_0	Toggle

↑ = clock transition LOW to HIGH
Q_0 = output level prior to clock transition

Table 8.7 Truth table for D flip-flop

INPUT		OUTPUT
D	*CLK*	*Q*
0	↑	0
0	↑	1

and registers. When we're done with individual flip-flops, we'll go on to counters and then look at registers.

Tables 8.4, 8.5, 8.6, 8.7 are the truth tables for type JK, RS, SC, D of flip-flop.

Given the waveforms in Fig. 8.7 for the *JK* input and the clock, determine the *Q* output waveform.

The D flip-flop is useful when a single data bit (1 or 0) is to be stored. The addition of an inverter to a J-K flip-flop creates a basic D flip-flop, as in Fig. 8.8.

Given the waveforms in Fig. 8.9 for the D input and the clock, determine the *Q* output waveform.

If the T input is HIGH, the T flip-flop changes state ("toggles") whenever the clock input is strobed. If the T input is LOW, the flip-flop holds the previous value. This behavior is described by the characteristic equation: $Q_{next} = Q{\cdot}\overline{T} + \overline{Q}{\cdot}T = Q \otimes T$.

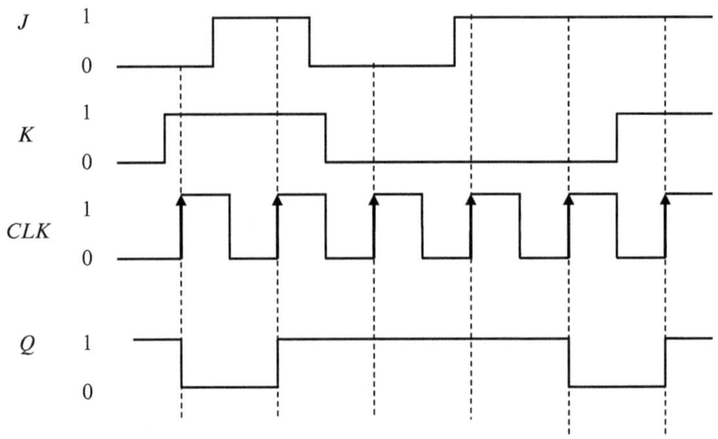

Fig. 8.7 The waveforms for the JK input and the clock

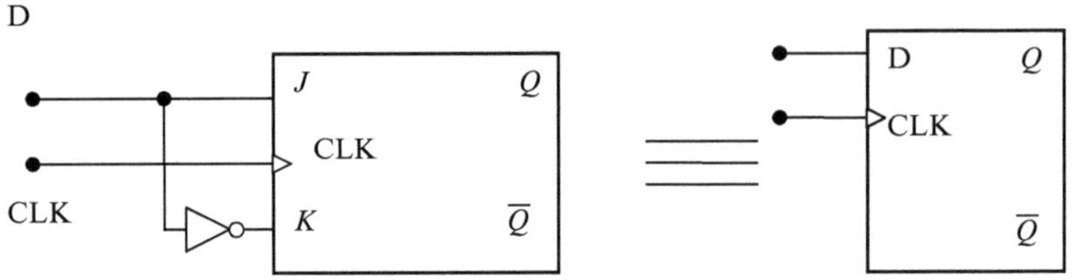

Fig. 8.8 The addition of an inverter to a J-K flip-flop creates a basic D flip-flop

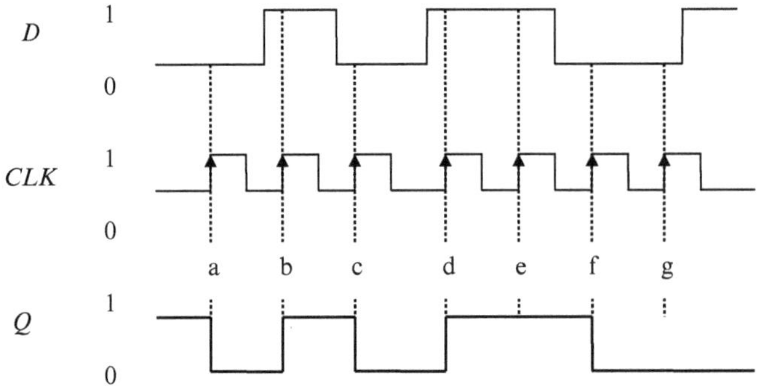

Fig. 8.9 The waveforms for the D input and the clock

It is useful for constructing binary counters, frequency dividers, and general binary addition devices. It can be made from a J-K flip-flop by tying both of its inputs HIGH (Table 8.8).

Construction of T flip-flop from a J-K flip-flop, Fig. 8.10.

Table 8.8 T flip-flop operation

Characteristic table				Excitation table			
T	Q	Q^{next}	Comment	T	Q	Q^{next}	Comment
0	0	0	Hold state (no clk)	0	0	0	No change
0	1	1	Hold state (no clk)	1	1	0	No change
1	0	1	Toggle	0	1	1	Complement
1	1	1	Toggle	1	0	1	Complement

Fig. 8.10 T - flip-flop

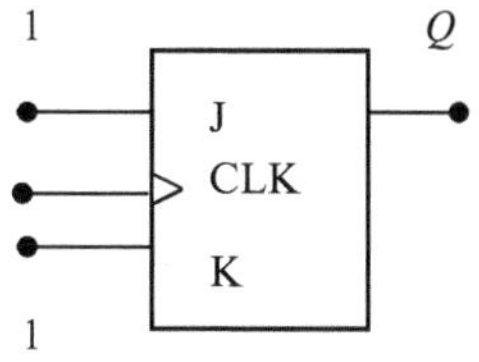

Fig. 8.11 Logic symbol for a J-K flip-flop with preset and clear inputs

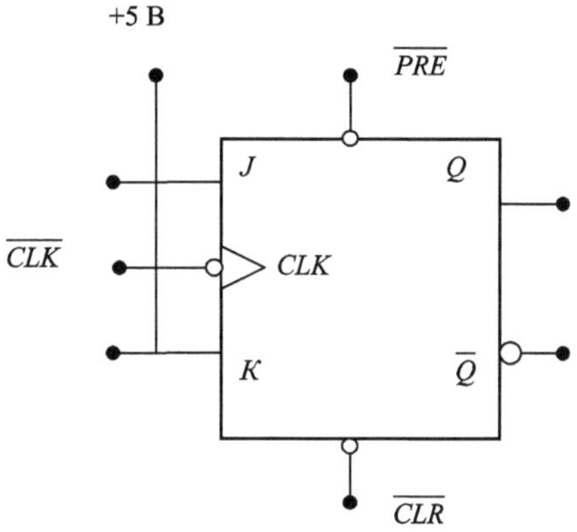

8.2.2 Asynchronous Preset and Clear Inputs

Most integrated circuit flip-flops also have asynchronous inputs. These are inputs that affect the state of the flip-flop independent of the clock. They are normally labeled preset (PRE) and clear (CLR), or direct set and direct reset by some manufacturers. An active level on the preset input will set the flip-flop, and an active level on the clear input will reset it. A logic symbol for a J-K flip-flop with preset and clear inputs is shown in Fig. 8.11. These inputs are active-LOW, as indicated by the bubbles. These preset and clear inputs must both be kept HIGH for synchronous operation (Table 8.9).

Asynchronous inputs on a flip-flop have control over the outputs (Q and $\overline{Q}$) regardless of clock input status.

Table 8.9 Truth table for D flip-flop

$\overline{\text{PRESET}}$	$\overline{\text{CLEAR}}$	COMMENTS
1	1	*Synchronized work*
0	1	$Q=1$ (*regardless of the CLK*)
1	0	$Q=0$ (*regardless of the CLK*)
0	0	*Not used*

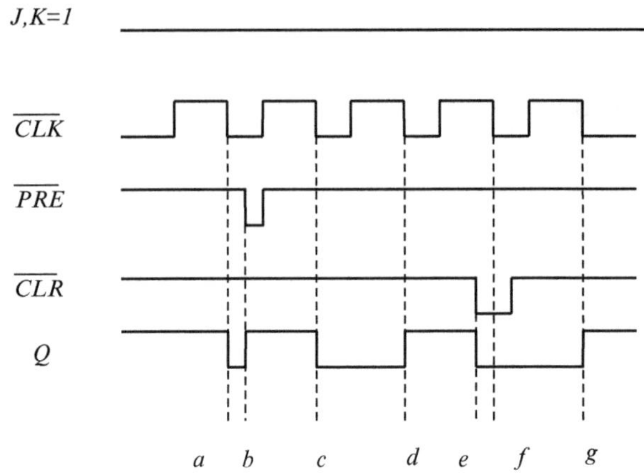

Fig. 8.12 Waveforms for the J-K

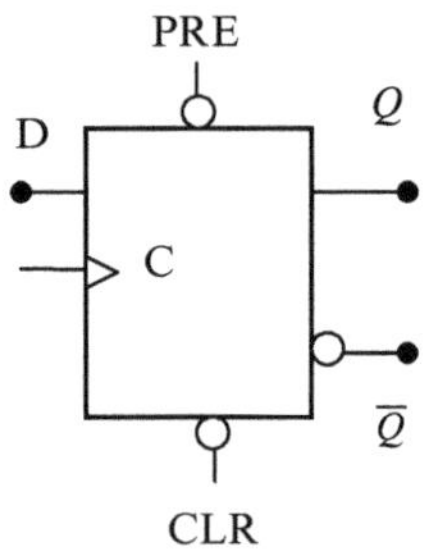

Fig. 8.13 Asynchronous flip-flop (D) inputs

These inputs are called the preset (PRE) and clear (CLR). The preset input drives the flip-flop to a set state while the clear input drives it to a reset state.

It is possible to drive the outputs of a J-K flip-flop to an invalid condition using the asynchronous inputs, because all feedback within the multivibrator circuit is overridden.

Given the waveforms $\overline{\text{PRESET}}$ and $\overline{\text{CLEAR}}$ in Fig. 8.12 for the J-K inputs and the clock, determine the Q output waveform.

Asynchronous inputs, just like synchronous inputs, can be engineered to be active-HIGH or active-LOW. If they're active-LOW, there will be an inverting bubble at that input lead on the block symbol, just like the negative edge-trigger clock inputs (Figs. 8.13, 8.14, 8.15).

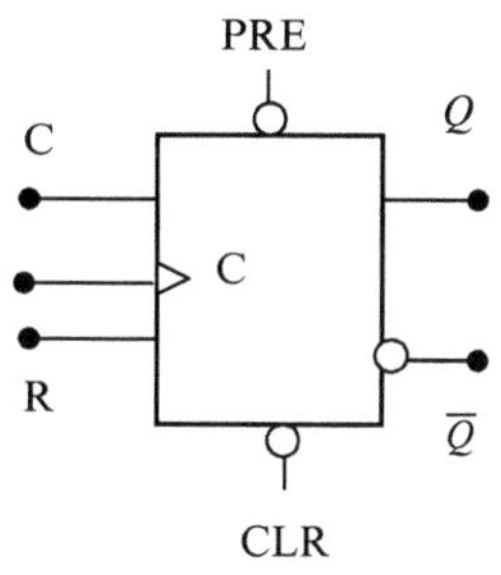

Fig. 8.14 Asynchronous flip-flop (C-R) inputs

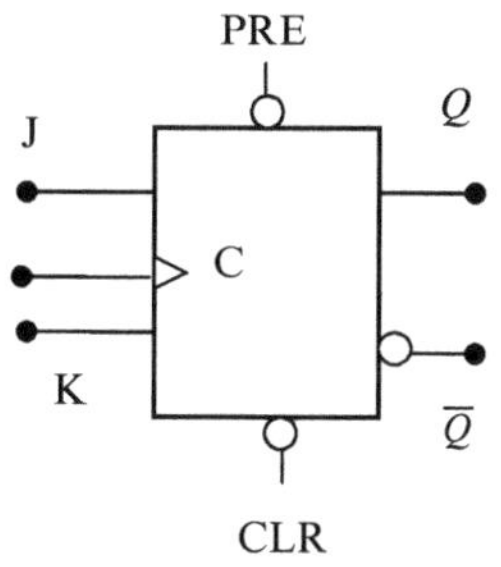

Fig. 8.15 Asynchronous flip-flop (J-K) inputs

8.3 Counters

In digital logic and computing, a counter is a device which stores (and sometimes displays) the number of times a particular event or process has occurred, often in relationship to a clock signal. In practice, there are two types of counters:

- Up counters, which increase (increment) in value;
- Down counters, which decrease (decrement) in value.

8.3.1 A Basic Digital Counter

One common requirement in digital circuits is counting, both forward and backward. The demonstration below shows the most basic kind of binary counting circuit.

Figure 8.16 shows a 4-bit counter. The output of each flip-flop changes state on the falling edge (1-to-0 transition) of the T input. The count held by this counter is read in the reverse order from the order in which the flip-flops are triggered. Thus, output *D* is the HIGH order of the count, while output A is the LOW order. The binary count held by the counter is then *DCBA*, and runs from 0000 (decimal 0) to 1111 (decimal 15). The next clock pulse will cause the counter to try to increment to 10000 (decimal 16). However, that 1 bit is not held by any flip-flop

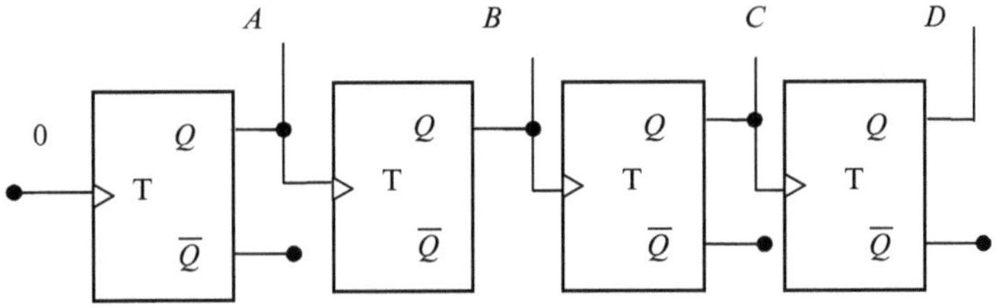

Fig. 8.16 4-bit counter

and is therefore lost. As a result, the counter actually reverts to 0000, and the count begins again.

A major problem with the counter (Fig. 8.16) is that the individual flip-flops do not all change state at the same time. Rather, each flip-flop is used to trigger the next one in the series.

8.3.2 Synchronous Counter

We noted the need to have all flip-flops in a counter to operate in unison with each other, so that all bits in the output count would change state at the same time. To accomplish this, we need to apply the same clock pulse to all flip-flops.

However, we do not want all flip-flops to change state with every clock pulse. Therefore, we'll need to add some controlling gates to determine when each flip-flop is allowed to change state, and when it is not. This requirement denies us the use of T flip-flops, but does require that we still use edge-triggered circuits. We can use either RS or JK flip-flops for this; we'll use JK flip-flops for the demonstrations on this section.

8.3.2.1 A Synchronous Binary Counter

This section begins our study of designing an important class of clocked sequential logic circuits-synchronous finite-state machines. Like all sequential circuits, a finite-state machine determines its outputs and its next state from its current inputs and current state. A synchronous finite-state machine changes state only on the clocking event.

A simple way of implementing the logic for each bit of an ascending counter is for each bit to toggle when all of the less significant bits are at a logic HIGH state. For example, bit 1 toggles when bit 0 is logic HIGH; bit 2 toggles when both bit 1 and bit 0 are logic HIGH; bit 3 toggles when bit 2, bit 1 and bit 0 are all HIGH; and so on. Synchronous counters can also be implemented with hardware that consists of finite state machines, which are more complex but allow for smoother, more stable transitions.

Table 8.10 Truth table for all states of the counter

D	C	B	A	Count
0	0	0	0	0
0	0	0	1	1
0	0	1	0	2
0	0	1	1	3
0	1	0	0	4
0	1	0	1	5
0	1	1	0	6
0	1	1	1	7
1	0	0	0	8
1	0	0	1	9
1	0	1	0	10
1	0	1	1	11
1	1	0	0	12
1	1	0	1	13
1	1	1	0	14
1	1	1	1	15

To determine the gates required at each flip-flop input, start by drawing up a truth table for all states of the counter (Table 8.10). Looking first at output A, we note that it must change state with every input clock pulse. But even with JK flip-flops, all we need to do here is to connect both the J and K inputs of this flip-flop to logic 1 in order to get the correct activity.

Flip-flop B is a bit more complicated. This output must change state only on every other input clock pulse. Looking at the truth table again, output B must be ready to change states whenever output A is a logic 1, but not when A is a logic 0. If we recall the behavior of the JK flip-flop, we can see that if we connect output A to the J and K inputs of flip-flop B, we will see output B behaving correctly.

Output *C* may change state only when both *A* and *B* are logic 1. We can't use only output *B* as the control for flip-flop *C*; that will allow *C* to change state when the counter is in state 2, causing it to switch directly from a count of 2 to a count of 7, and again from a count of 10 to a count of 15 – not a good way to count. Therefore we will need a two-input AND gate at the inputs to flip-flop *C*. Flip-flop *D* requires a three-input AND gate for its control, as outputs *A*, *B*, and *C* must all be at logic 1 before *D* can be allowed to change state. The resulting circuit is shown in the demonstration below, Fig. 8.17.

8.3.3 Decimal and Shorter Counts

Example 1

To create a counter ($M=5$), we need to find a way to cut the counting sequence short. The truth table to the left shows the actual counting sequence we need. Note that the counting sequence is exactly the same as for the binary counter. At that

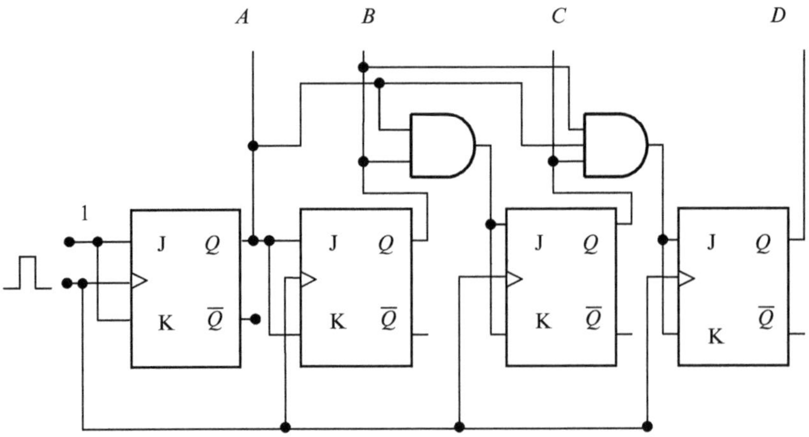

Fig. 8.17 The scheme of synchronous binary counter

Fig. 8.18 A state diagram for a counter

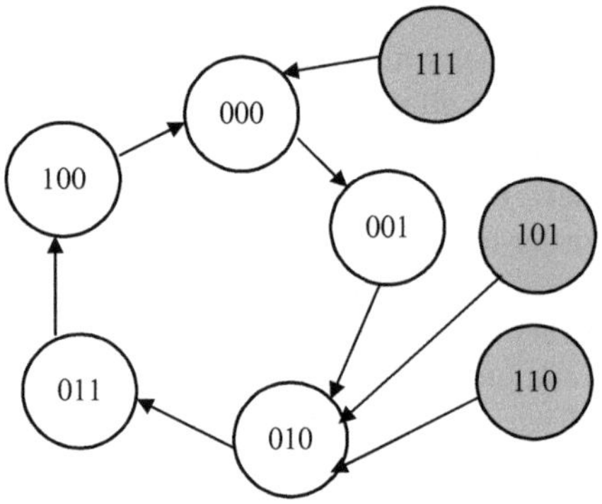

point, where the binary counter would continue on to a count of 4, the counter must reset itself to a count of 0.

Step 1: State diagram

The first step in the design of a counter is to create a state diagram. A state diagram shows the progression of states through which the counter advances when it is clocked. As an example, Fig. 8.18 is a state diagram for a counter.

Step 2: Next-state table and Step 3: Flip-flop transition table

Once the sequential circuit is defined by a state diagram, the second step is to derive a next-state table, which lists each state of the counter (present state) along with the corresponding next state. The next state is the state that the counter goes to from its present state upon application of a clock pulse. The next-state table is derived from the state diagram and is shown in Tables 8.11 and 8.12. Once the state diagram of the sequential circuit is defined, a Next - State Table is derived which lists each present state and the corresponding next state. The next state is the state to which the sequential circuit switches when a clock transition occurs.

Table 8.11 Transition table for a J-K flip-flop

Output transitions	Present state $Q_{(N)}$	Next state $Q_{(N+1)}$	J	K
0-0	0	0	0	x
0-1	0	1	1	x
1-0	1	0	x	1
1-1	1	1	x	0

Table 8.12 JK flip-flop truth table

Present state			Next state								
Q_2	Q_1	Q_0	Q_2	Q_1	Q_0	J_2	K_2	J_1	K_1	J_0	K_0
0	0	0	0	0	1	0	x	0	x	1	x
0	0	1	0	1	0	0	x	1	x	x	1
0	1	0	0	1	1	0	x	x	0	1	x
0	1	1	1	0	0	1	x	x	1	x	1
1	0	0	0	0	0	x	1	0	x	0	x

Step 4: Minimization and Step 5: Logic expressions for flip-flop inputs

$$J_2 = Q_1 Q_0.$$

$$J_1 = Q_0.$$

$$J_0 = \overline{Q}_2.$$

$$K_2 = K_0 = 1.$$

$$K_1 = J_1.$$

Step 6: Counter implementation
The Boolean expressions obtained in the previous step are implemented using logic gates. The sequential circuit implemented is shown in Fig. 8.19. The hardware diagram of the counter (Fig. 8.19).
Example 2

Step 1: State diagram (Fig. 8.20)

Step 2: Next-state table and Step 3: Flip-flop transition table (Table 8.13)

Step 4: Minimization and Step 5: Logic expressions for flip-flop inputs
The following diagram shows the steps to create separate next states of separate J and K from the current states of J and K (Figs. 8.21 and 8.22).

Step 6: Counter implementation
The circuit of the counter (Figs. 8.23 and 8.24).
The process of implementation of the counter design is shown in Appendix C.
In a digital circuit, an FSM may be built using a programmable logic device, a programmable logic controller, logic gates and flip-flops or relays. More specifically,

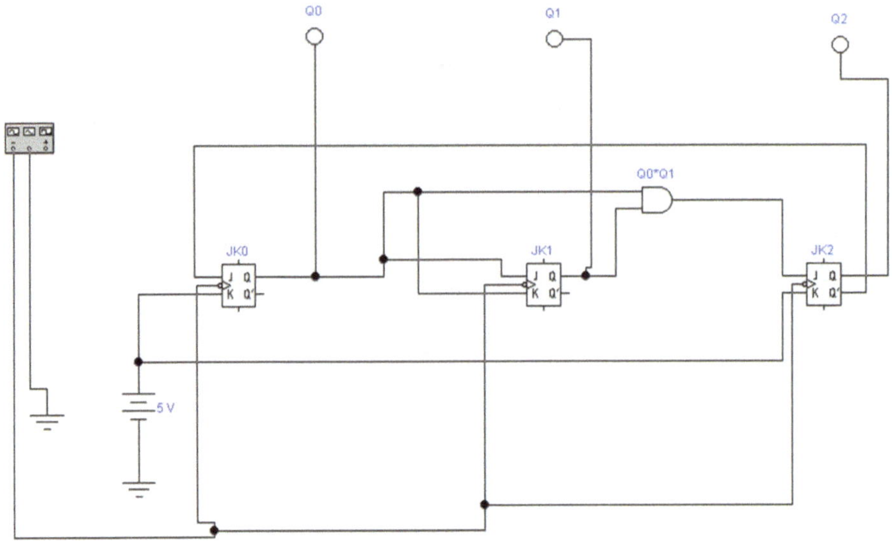

Fig. 8.19 A synchronous counter using JK flip-flops

Fig. 8.20 A state diagram for a counter

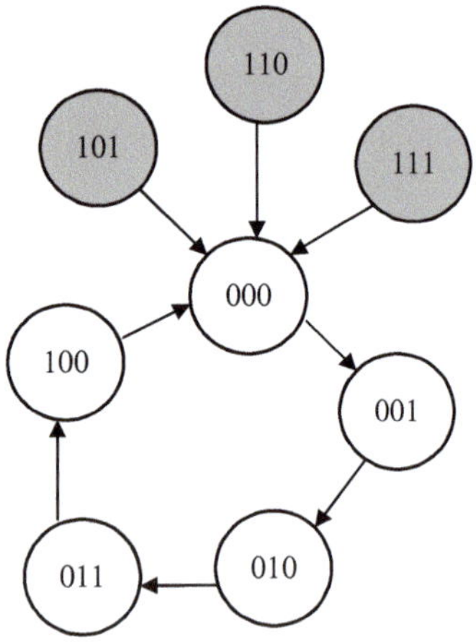

Table 8.13 JK flip-flop state table

Present state			Next state								
C	B	A	C	B	A	J_C	K_C	J	K	J_A	K_A
0	0	0	0	0	1	0	x	0	x	1	x
0	0	1	0	1	0	0	x	1	x	x	1
0	1	0	0	1	1	0	x	x	0	1	x
0	1	1	1	0	0	1	x	x	1	x	1
1	0	0	0	0	0	x	1	0	x	0	x
1	0	1	0	0	0	x	1	0	x	x	1
1	1	0	0	0	0	x	1	x	1	0	x
1	1	1	0	0	0	x	1	x	1	x	1

PRESENT STATE			J_A
C	B	A	
0	0	0	1
0	0	1	x
0	1	0	1
0	1	1	x
1	0	0	0
1	0	1	x
1	1	0	0
1	1	1	x

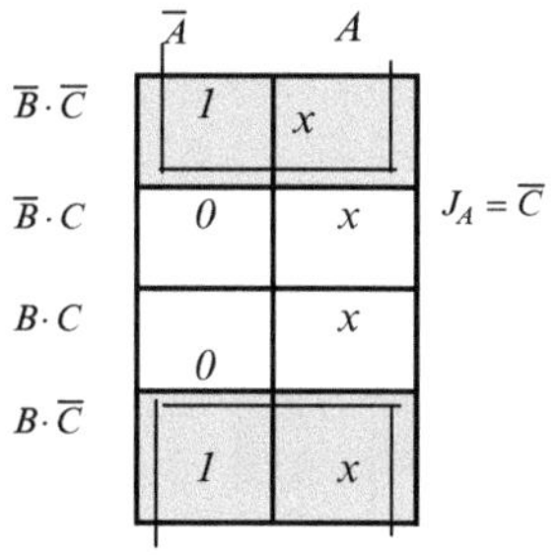

Fig. 8.21 Minimization of logic expressions for flip-flop input J_A

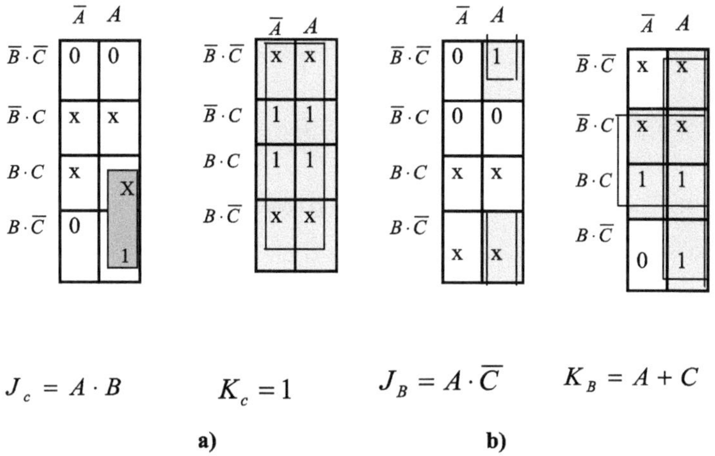

a) for J_C and K_C; b) for J_B and K_B

Fig. 8.22 Minimization of logic expressions for flip-flop inputs $J_C\,K_C$; J_B and K_B

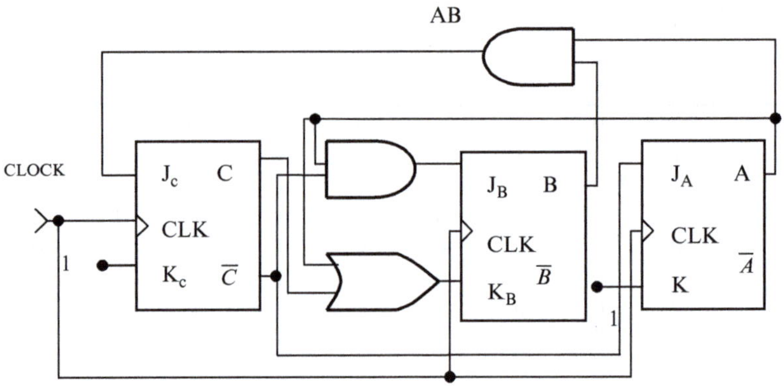

Fig. 8.23 The circuit of the counter

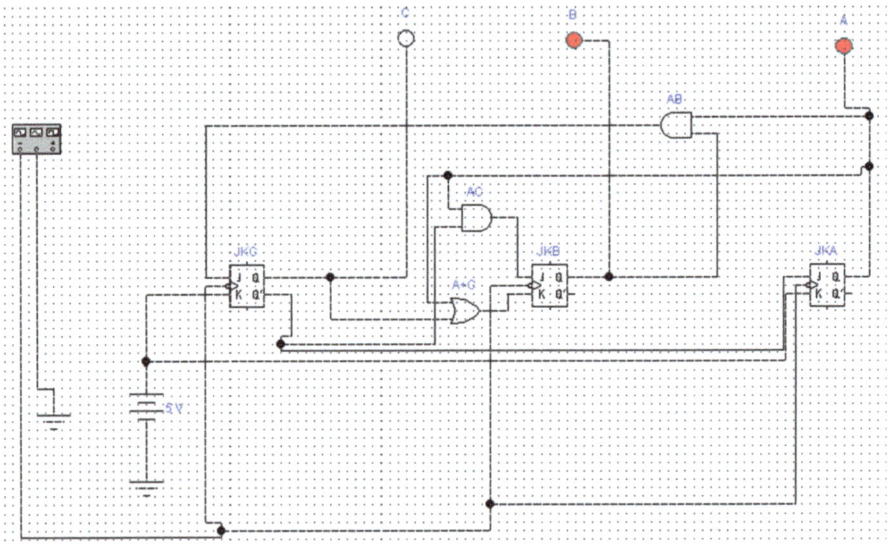

Fig. 8.24 The circuit of the counter (Electronic Workbench 5.12)

a hardware implementation requires a register to store state variables, a block of combinational logic which determines the state transition, and a second block of combinational logic that determines the output of an FSM.

Consider an example in which you want to design a meter that works with the expense ratio 7, when the input signal $X=1$, and the expense ratio 5, when the input signal $X=0$.

Counters will be designed using the T-flip-flops or J-K flip-flops. The data for the design of the counters are shown in Tables 8.14 and 8.15.

Figures 8.25, 8.26, 8.27 show Karnaugh Maps. Figure 8.28 – circuit of the counter.-

Table 8.14 The data for the design of the counters

		Present state						Next state flip-flop		
		Present			Next			T		
No.	X	Q_2	Q_1	Q_0	Q_2^+	Q_1^+	Q_0^+	T_2	T_1	T_0
0	0	0	0	0	0	0	1	0	0	1
1	0	0	0	1	0	1	0	0	1	1
2	0	0	1	0	0	1	1	0	0	1
3	0	0	1	1	1	0	0	1	1	1
4	0	1	0	0	0	0	0	1	0	0
5	0	1	0	1	*	*	*	*	*	*
6	0	1	1	0	*	*	*	*	*	*
7	0	1	1	1	*	*	*	*	*	*
8	1	0	0	0	0	0	1	0	0	1
9	1	0	0	1	0	1	0	0	1	1
10	1	0	1	0	0	1	1	0	0	1
11	1	0	1	1	1	0	0	1	1	1
12	1	1	0	0	1	0	1	0	0	1
13	1	1	0	1	1	1	0	0	1	1
14	1	1	1	0	0	0	0	1	1	0
15	1	1	1	1	*	*	*	*	*	*

Table 8.15 The data for the design of the counters

		Present state						Next state flip-flop					
		Present			Next			JK					
No.	X	Q_2	Q_1	Q_0	Q_2^+	Q_1^+	Q_0^+	J_2	K_2	J_1	K_1	J_0	K_0
0	0	0	0	0	0	0	1	0	*	0	*	1	*
1	0	0	0	1	0	1	0	0	*	1	*	*	1
2	0	0	1	0	0	1	1	0	*	*	0	1	*
3	0	0	1	1	1	0	0	1	*	*	1	*	1
4	0	1	0	0	0	0	0	*	1	0	*	0	*
5	0	1	0	1	*	*	*	*	*	*	*	*	*
6	0	1	1	0	*	*	*	*	*	*	*	*	*
7	0	1	1	1	*	*	*	*	*	*	*	*	*
8	1	0	0	0	0	0	1	0	*	0	*	1	*
9	1	0	0	1	0	1	0	0	*	1	*	*	1
10	1	0	1	0	0	1	1	0	*	*	0	1	*
11	1	0	1	1	1	0	0	1	*	*	1	*	1
12	1	1	0	0	1	0	1	*	0	0	*	1	*
13	1	1	0	1	1	1	0	*	0	1	*	*	1
14	1	1	1	0	0	0	0	*	0	1	*	*	1
15	1	1	1	1	*	*	*	*	*	*	*	*	*

Fig. 8.25 Karnaugh maps for T_2

For T_2.

xQ_2 \ Q_1Q_0	00	01	11	10
00	0	0	1	0
01	1	*	*	*
11	0	0	*	1
10	0	0	1	0

$$T_2 = Q_1 \cdot Q_2 \vee \overline{x} \cdot Q_2 \vee Q_1 \cdot Q_0$$

Fig. 8.26 Karnaugh maps for T_1

For T_1.

xQ_2 \ Q_1Q_0	00	01	11	10
00	0	1	1	0
01	0	*	*	*
11	0	1	*	1
10	0	1	1	0

$$T_1 = Q_0 \vee Q_2 \cdot Q_1$$

Fig. 8.27 Karnaugh maps for T_0

For T_0.

xQ_2 \ Q_1Q_0	00	01	11	10
00	1	1	1	1
01	0	*	*	*
11	1	1	*	1
10	1	1	1	1

$$T_0 = \overline{Q_2} \vee x \cdot \overline{Q_1}$$

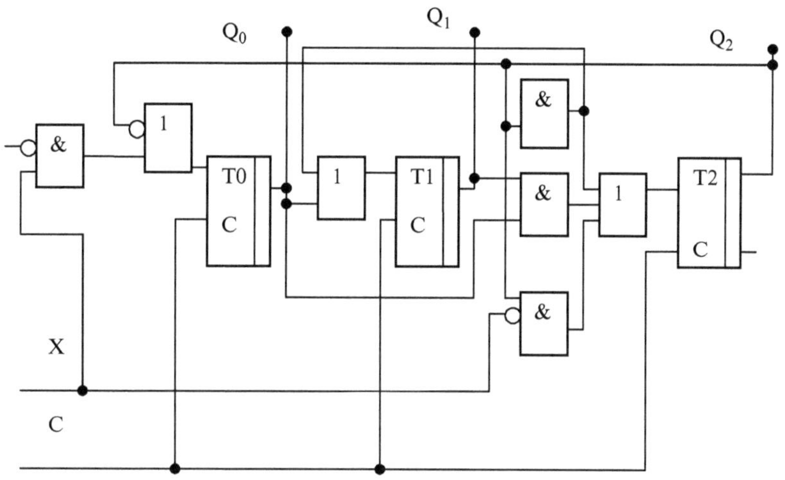

Fig. 8.28 Implementation of the sequential circuit

For T_2.
For ***JK***

$$\begin{cases} J_2 = Q_1 \cdot Q_2; \\ K_2 = \overline{X} + Q_1; \\ J_1 = Q_0; \\ K_1 = Q_2 + Q_0; \\ J_0 \equiv \overline{Q}_2 + X \cdot \overline{Q}_1; \\ K_0 = 1. \end{cases}$$

T flip-flop is a model of JK or D trigger, so this scheme is fundamental and cannot be collected in the EWB 5.12 or Multisim.

Examples

Example 1. Figure 8.29 shows a circuit of the counter, that seven-segment displays the numbers from 0 to 5, when the input on the key of $X=0$ and from 0 to 12, when $X=1$.

Example 2. Figure 8.30 shows a circuit of the counter, that seven-segment displays the numbers from 0 to 15, when the input on the key of $X=0$ and from 0 to 22, when $X=1$. Counters with variable module accounts are used primarily as a frequency dividers with adjustable coefficient. Such circuits are used for example in radio frequency technology, where there exists continual demand for the development of circuits with ever higher clock rates or frequencies. In order to realize frequency divider circuits, usually a plurality of gates are connected in series in a combinatorial part of the circuit, so that, for each state change of the input signal, many gates are switched within one clock period.

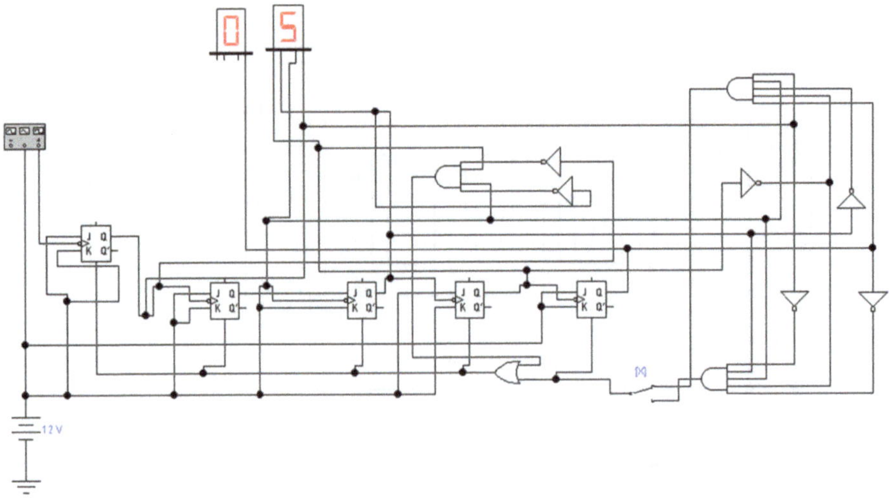

Fig. 8.29 The circuit of the counter that seven-segment displays the numbers from 0 to 5 (Electronic Workbench 5.12)

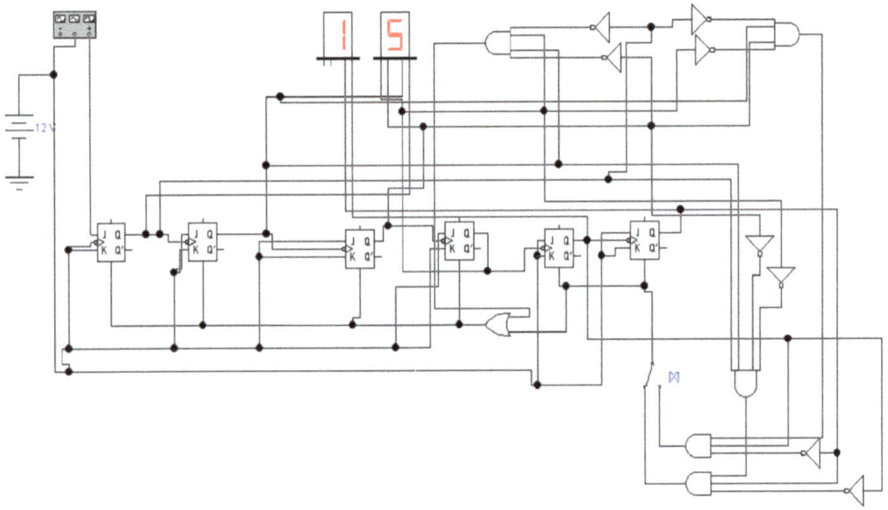

Fig. 8.30 The circuit of the counter that seven-segment displays the numbers from 0 to 15 (Electronic Workbench 5.12)

8.3.4 BCD Counter

Binary-coded-decimal (BCD) counters can be designed using the approach explained in Chap. 3. A two-digit BCD counter is presented in Fig. 8.31 (a two-digit BCD counter). It consists of two modulo-10 counters, one for each BCD digit, which we

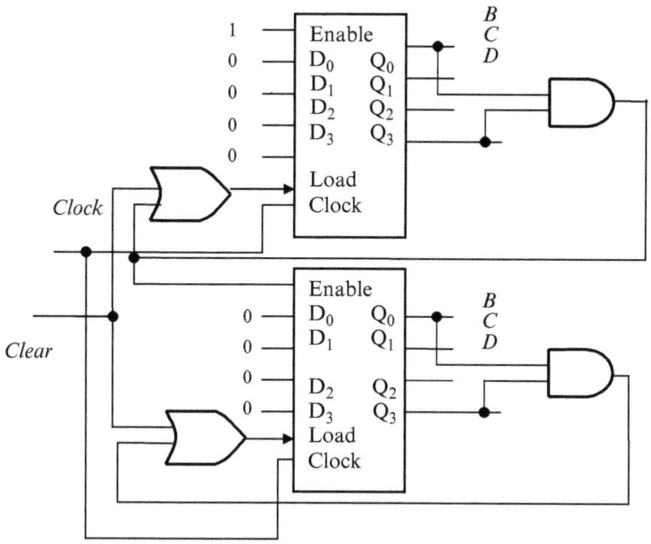

Fig. 8.31 BCD Counter

implemented using the parallel-load four-bit counter. Note that in a modulo-10 counter it is necessary to reset the four flip-flops after the count of 9 has been obtained. Thus the Load input to each stage is equal to 1 when $Q_3 = Q_0 = 1$, which causes 0 s to be loaded into the flip-flops at the next positive edge of the clock signal. Whenever the count in stage 0, $BCD_{(0)}$, reaches 9 it is necessary to enable the second stage so that it will be incremented when the next clock pulse arrives. This is accomplished by keeping the Enable signal for $BCD_{(1)}$ LOW at all times except when $BCD_{(0)} = 9$.

In practice, it has to be possible to clear the contents of the counter by activating some control signal. Two OR gates are included in the circuit for this purpose. The control input *Clear* can be used to load 0 s into the counter. Observe that in this case *Clear* is active when HIGH. In any digital system there is usually one or more clock signals used to drive all synchronous circuitry. In the preceding counter, as well as in all counters presented in the previous figures, we have assumed that the objective is to count the number of clock pulses. Of course, these counters can be used to count the number of pulses in any signal that may be used in place of the clock signal.

8.3.5 *The Johnson Counter*

In some cases, we want a counter that provides individual digit outputs rather than a binary or BCD output. Of course, we can do this by adding a decoder circuit to the binary counter. However, in many cases it is much simpler to use a different counter structure, that will permit much simpler decoding of individual digit outputs.

Table 8.16 Truth table of Johnson counter

States					
A	*B*	*C*	*D*	*E*	Count
0	0	0	0	0	0
1	0	0	0	0	1
1	1	0	0	0	2
1	1	1	0	0	3
1	1	1	1	0	4
1	1	1	1	1	5
0	1	1	1	1	6
0	1	1	1	1	7
0	0	0	1	1	8
0	0	0	0	1	9

ww

For example, consider the counting sequence, Table 8.16. It actually resembles the behavior of a shift register more than a counter, but that need not be a problem. Indeed, we can easily use a shift register to implement such a counter. In addition, we can notice that each legal count may be defined by the location of the last flip-flop to change states, and which way it changed state. This can be accomplished with a simple two-input AND or NOR gate monitoring the output states of two adjacent flip-flops. In this way, we can use ten simple 2-input gates to provide ten decoded outputs for digits 0–9. This is known as the Johnson counting sequence, and counters that implement this approach are called Johnson Counters.

Johnson Ring Counters or "Twisted Ring Counters", are exactly the same idea as the Walking Ring Counter above, except that the inverted output Q of the last Flip-flop is connected back to the input D of the first Flip-flop as shown below. The main advantage of this type of ring counter is that it only needs half the number of Flip-flops compared to the standard walking ring counter in which its Modulo number is halved.

This inversion of Q before it is fed back to input **D** causes the counter to "count" in a different way. Instead of counting through a fixed set of patterns like the walking ring counter such as for a 4-bit counter, "1000"(1), "0100"(2), "0010"(4), "0001"(8) etc., the Johnson counter counts up and then down as the initial logic "1" passes through it to the right replacing the preceding logic "0". A 4-bit Johnson ring counter passes blocks of four logic "0" and then four logic "1" thereby producing an 8-bit pattern. As the inverted output $\overline{Q}_4$ is connected to the input **D** this 8-bit pattern continually repeats. For example, "1000", "1100", "1110", "1111", "0111", "0011", "0001", "0000" and this is demonstrated in Table 8.17 and Fig. 8.32.

As well as counting, Ring Counters can be used to detect or recognise various patterns or number values. By connecting simple logic gates such as AND or OR gates to the outputs of the Flip-flops, the circuit can be made to detect a set number or value. Standard 2, 3 or 4-stage Johnson Ring Counters can also be used to divide the frequency of the clock signal by varying their feedback connections, and divide-by-3 or divide-by-5 outputs are also available.

Table 8.17 Truth table for a 4-bit Johnson ring counter

*D*1	*D*2	*D*3	*D*4
0	0	0	0
1	0	0	0
1	1	0	0
1	1	1	0
1	1	1	1
0	1	1	1
0	0	1	1
0	0	0	1

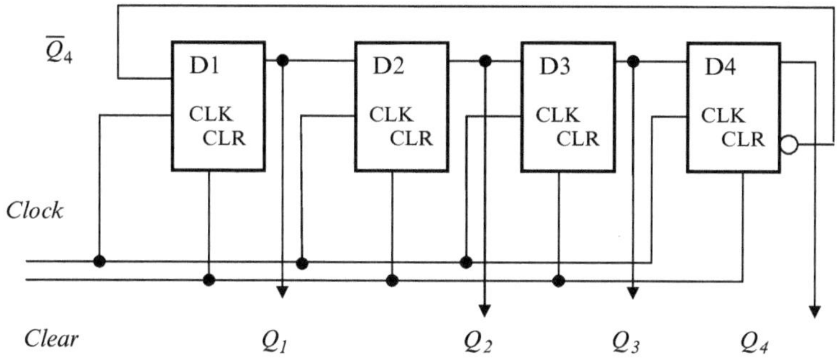

Fig. 8.32 4-bit Johnson Ring Counter

8.4 Registers

A Shift Register consists of a number of single bit "D-Type Data Latches" connected together in a chain arrangement so that the output from one data latch becomes the input of the next latch and so on, thereby moving the stored data serially from either the left or the right direction. The number of individual Data Latches used to make up a Shift Register is determined by the number of bits to be stored with the most common being 8-bits wide. Shift Registers are mainly used to store data and to convert data from either a serial to parallel or parallel to serial format with all the latches being driven by a common clock (*CLK*) signal making them Synchronous devices. They are generally provided with a *Clear* or *Reset* connection so that they can be "*SET*" or "*RESET*" as required.

Generally, Shift Registers operate in one of four different modes:

- Serial-in to Parallel-out (SIPO);
- Serial-in to Serial-out (SISO);
- Parallel-in to Parallel-out (PIPO);
- Parallel-in to Serial-out (PISO).

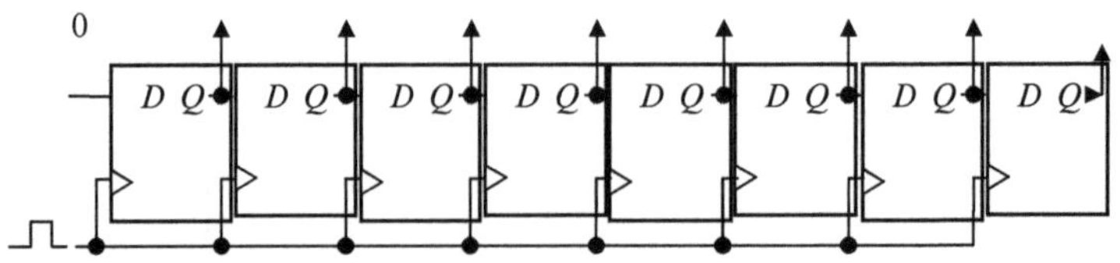

Fig. 8.33 Serial-to-Parallel Shift Register

In this section, we consider two types of registers – Serial-to-Parallel Shift Register and Parallel-to-Serial Shift Register.

8.4.1 Serial-to-Parallel Shift Register

The term register can be used in a variety of specific applications, but in all cases it refers to a group of flip-flops operating as a coherent unit to hold data. This is different from a counter, which is a group of flip-flops operating to generate new data by tabulating it.

A counter can be viewed as a specialized kind of register, which counts events and thereby generates data, rather than just holding the data or changing the way it is handled. The demonstration circuit below is known as a shift register because data is shifted through it, from flip-flop to flip-flop. If you apply one byte (8 bits) of data to the initial data input one bit at a time, and apply one clock pulse to the circuit after setting each bit of data, you will find the entire byte present at the flip-flop outputs in parallel format. Therefore, this circuit is known as a serial-in, parallel-out shift register. It is also known sometimes as a shift-in register, or as a serial-to-parallel shift register. By standardized convention, the least significant bit (LSB) of the byte is shifted in first (Fig. 8.33).

8.4.2 Parallel-to-Serial Shift Register

Where there is a need for serial-to-parallel conversion, there is also a need for parallel-to-serial conversion. The parallel-in, serial-out register (or parallel-to-serial shift register, or shift-out register). Since each flip-flop in the register must be able to accept data from either a serial or a parallel source, a small two-input multiplexer is required in front of each input. An extra input line selects between serial and parallel input signals, and as usual the flip-flops are loaded in accordance with a common clock signal.

A 4-bit shift register with parallel and serial inputs and outputs will fit nicely into a 14-pin DIP IC.

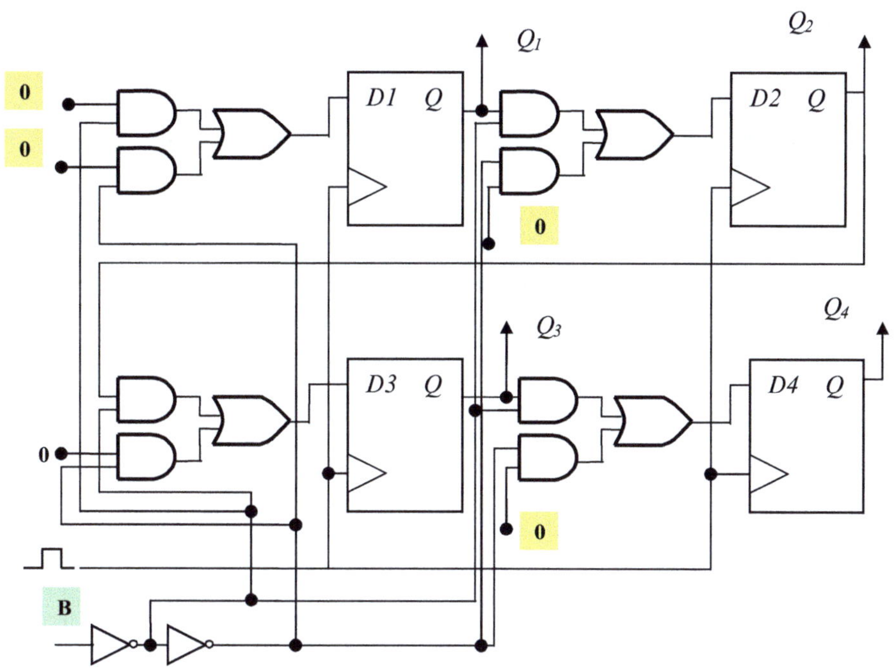

Fig. 8.34 Parallel-to-Serial Shift Register

The label "**B**" (for example button) indicates that the shift-out register is currently in serial mode. Thus, input signals present at the serial input just above the "**B**" label (button) will be shifted into the register one by one with each clock pulse. This enables us to load the entire register at once from the parallel inputs just below the multiplexers. Thus, we can have a parallel input and a serial output. The inclusion of a serial input makes it possible to cascade multiple circuits of this type in order to increase the number of bits in the total register.

Because this circuit has both parallel and serial inputs and outputs, it can serve as either a shift-in register or a shift-out register. This capability can have advantages in many cases (Fig. 8.34).

8.4.3 Using a Shift Register for Control

There are many ways to design a suitable control circuit for the swap operation. One possibility is to use the left-to-right shift register shown in Fig. 8.35. Assume that the reset input is used to clear the flip-flops to 0. Hence the control signals $R1_{in}$, $R1_{out}$, and so on are not asserted, because the shift register outputs have the value 0. The serial input w normally has the value 0. We assume that changes in the value of

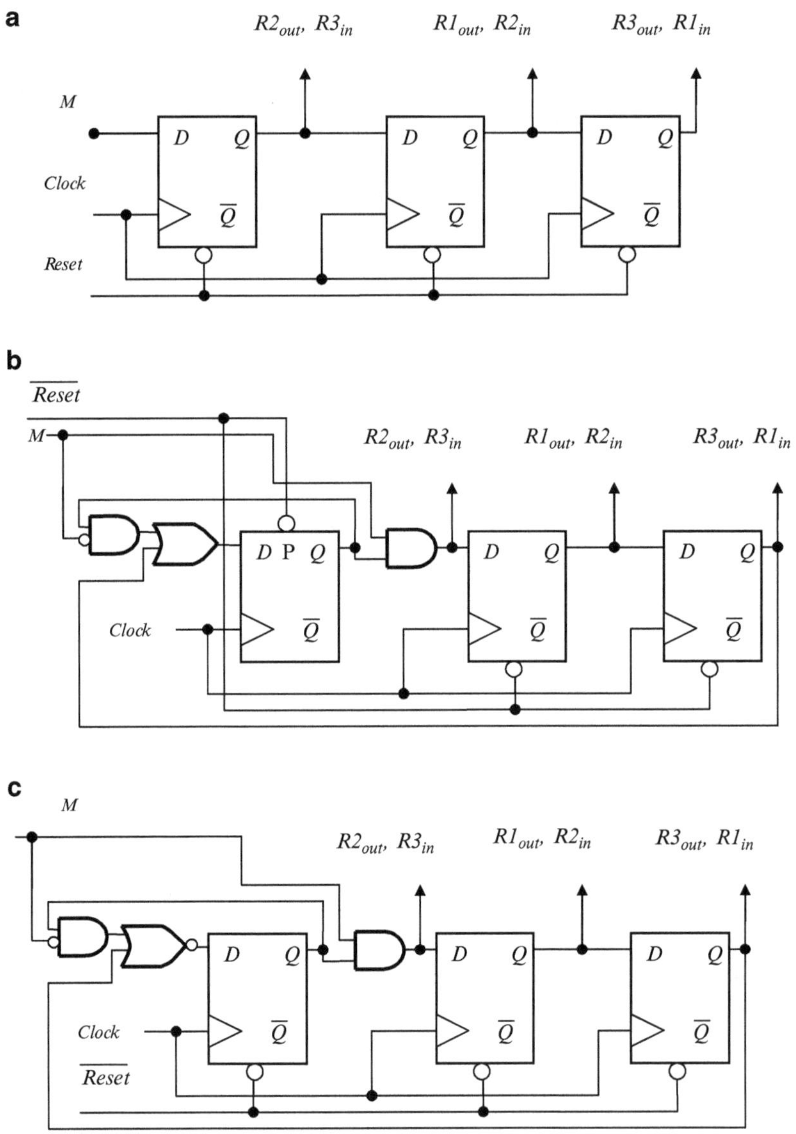

Fig. 8.35 The left-to-right shift register

M are synchronized to occur shortly after the active clock edge. This assumption is reasonable because *M* would normally be generated as the output of some circuit that is controlled by the same clock signal. When the desired swap should be performed, *M* is set to 1 for one clock cycle, and then *M* returns to 0. After the next active clock edge, the output of the left-most flip-flop becomes equal to 1, which asserts both $R2_{out}$ and $R3_{in}$. The contents of register *R2* are placed onto the bus wires and are loaded into register *R3* on the next active clock edge.

This clock edge also shifts the contents of the shift register, resulting in $R1_{out}=R2_{in}=1$. Note that since w is now 0, the first flip-flop is cleared, causing $R2_{out}=R3_{in}=0$. The contents of *R1* are now on the bus and are loaded into *R2* on the next clock edge. After this clock edge the shift register contains **001** and thus asserts $R3_{out}$ and $R1_{in}$. The contents of *R3* are now on the bus and are loaded into *R1* on the next clock edge. Using the control circuit in Fig. 8.35a, when *M* changes to 1 the swap operation does not begin until after the next active clock edge. We can modify the control circuit so that it starts the swap operation in the same clock cycle in which w changes to 1. One possible approach is illustrated in Fig. 8.35b. The reset signal is used to set the shift-register contents to **100**, by presetting the left-most flip-flop to 1 and clearing the other two flip-flops. As long as $M=0$, the output control signals are not asserted. When *M* changes to 1, the signals $R2_{out}$ and $R3_{in}$ are immediately asserted and the contents of *R2* are placed onto the bus. The next active clock edge loads this data into *R3* and also shifts the shift register contents to **010**. Since the signal $R1_{out}$ is now asserted, the contents of *R1* appear on the bus. The next clock edge loads this data into *R2* and changes the shift register contents to **001**. The contents of *R3* are now on the bus; this data is loaded into *R1* at the next clock edge, which also changes the shift register contents to **100**. We assume that w had the value 1 for only one clock cycle; hence the output control signals are not asserted at this point.

It may not be obvious to design a circuit such as the one in Fig. 8.35b, because we have presented the design in an ad hoc fashion. The circuit in Fig. 8.35b assumes that a preset input is available on the left-most flip-flop. If the flip-flop has only a clear input, then we can use the equivalent circuit shown in Fig. 8.35c. In this circuit we use the *Q* output of the left-most flip-flop and also complement the input to this flip-flop by using a NOR gate instead of an OR gate.

8.5 Timer

The 8-pin 555 timer IC is used in many projects. The 555 timer IC is an amazingly simple yet versatile device. It has been around now for many years and has been reworked into a number of different technologies. The two primary versions today are the original bipolar design and the more recent CMOS equivalent. These differences primarily affect the amount of power they require and their maximum frequency of operation; they are pin-compatible and functionally interchangeable.

The figure to the right shows the functional block diagram of the 555 timer IC. The IC is available in either an 8-pin round TO3-style can or an 8-pin mini-DIP package. In either case, the pin connections are as follows:

The operation of the 555 timer revolves around the three resistors that form a voltage divider across the power supply, and the two comparators connected to this voltage divider. The IC is quiescent so long as the trigger input (pin 2) remains at $+V_{CC}$ and the threshold input (pin 6) is at ground. Assume the reset input (pin 4) is also at $+V_{CC}$ and therefore inactive, and that the control voltage input (pin 5)

Fig. 8.36 The 555 timer IC

1. Ground.
2. Trigger input.
3. Output.
4. Reset input.
5. Control voltage.
6. Threshold input.
7. Discharge.
8. $+V_{CC.}$ +5 to +15 volts in normal use.

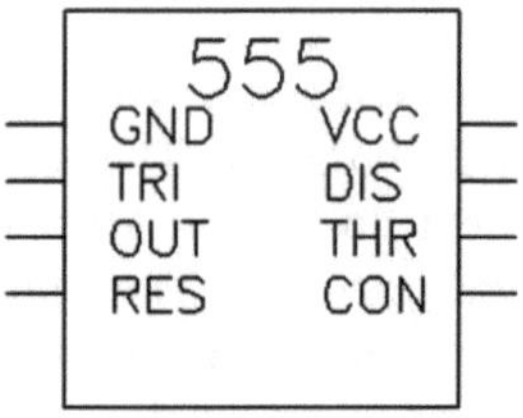

is unconnected. Under these conditions, the output (pin 3) is at ground and the discharge transistor (pin 7) is turned on, thus grounding whatever is connected to this pin.

The 555 can operate in either monostable or astable mode, depending on the connections to and the arrangement of the external components, Fig. 8.36. Thus, it can either produce a single pulse when triggered, or it can produce a continuous pulse train as long as it remains powered.

Figure 8.37 shows the use of the IC 555 (Astable Multivibrator). To get an accurate simulation, the number of Points per Cycle is set to 1000 on the Analysis Options dialog box under the Circuit menu and the analysis type is set to "Transient".

An astable circuit produces a "square wave", which is a digital waveform with sharp transitions between LOW (0 V) and HIGH ($+V_s$). Note that the durations of the LOW and HIGH states may be different. The circuit is called astable because it is not stable in any state: the output is continually changing between "LOW" and "HIGH".

The 556 is a dual version of the 555 housed in a 14-pin package; the two timers (A and B) share the same power supply pins.

Low power versions of the 555 are made, such as the ICM7555, but these should only be used when specified (to increase battery life) because their maximum output current of about 20 mA (with a *9 V* supply) is too low for many standard 555 circuits. The ICM7555 has the same pin arrangement as a standard 555.

One interesting and very useful feature of the 555 timer in either mode is that the timing interval for either charge or discharge is independent of the supply voltage, V_{CC}. This is because the same V_{CC} is used both as the charging voltage and as the basis of the reference voltages for the two comparators inside the 555.

With just a few external components IC 555 it can be used to build many circuits, Fig. 8.38.

This circuit demonstrates the use of the IC 555 timer in a monostable configuration.

A monostable circuit produces a single output pulse when triggered. It is called monostable because it is stable in just one state: "output LOW". The "output HIGH" state is temporary, Fig. 8.39.

The timing period is triggered (started) when the trigger input (555 pin 2) is less than $1/3V_{(in)}$, this makes the output HIGH $+V_{(in)}$ and the capacitor starts to charge through a resistor. Once the time period has started, further trigger pulses are ignored.

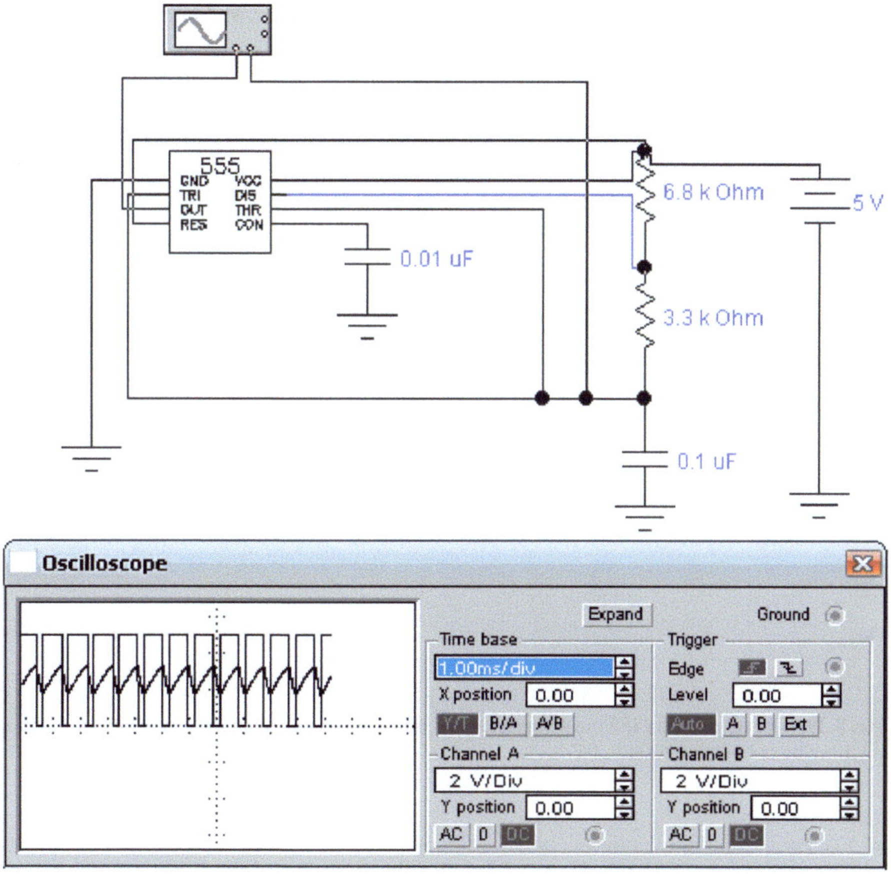

Fig. 8.37 Modeling of the timer (use the IC 555 - Astable Multivibrator)

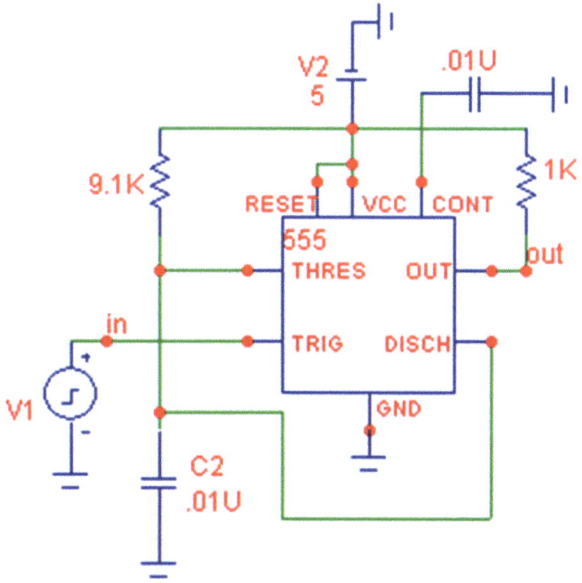

Fig. 8.38 An example of using a timer in a digital system

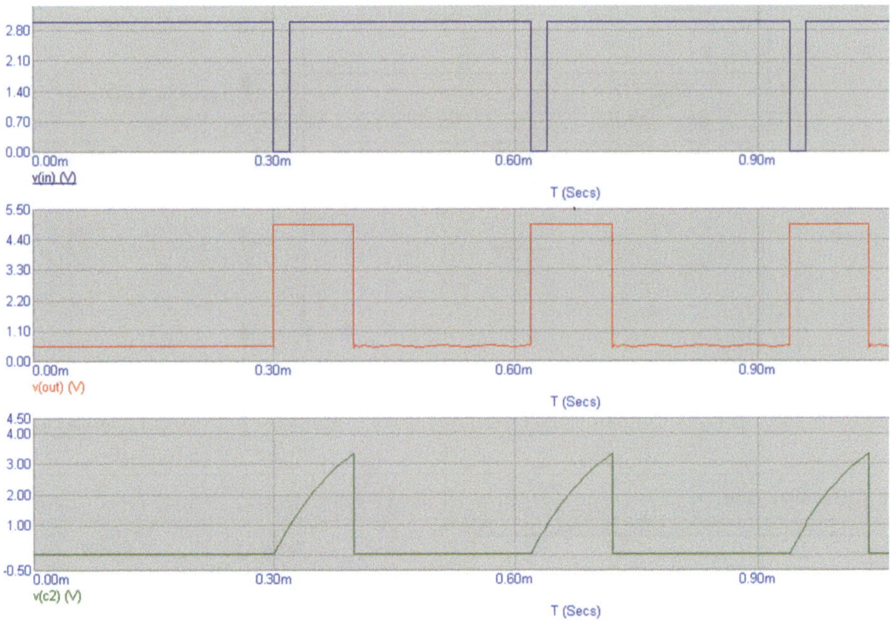

Fig. 8.39 The reset input (555 pin 4) overrides all other inputs and the timing may be cancelled at any time by connecting reset to 0V

The reset input (555 pin 4) overrides all other inputs and the timing may be cancelled at any time by connecting reset to 0 V, which instantly makes the output LOW and discharges the capacitor. If the reset function is not required the reset pin should be connected to $+V_{(in)}$.

8.6 Multiplexer and Demultiplexer

8.6.1 The Multiplexer

Data selectors, more commonly called a Multiplexer, shortened to "**Mux**" or "**MPX**", are combinational logic switching devices that operate like a very fast acting multiple position rotary switch. They connect or control multiple input lines called "channels" consisting of either 2, 4, 8 or 16 individual inputs, one at a time to an output. Then the job of a multiplexer is to allow multiple signals to share a single common output. For example, a single 8-channel multiplexer would connect one of its eight inputs to the single data output. Multiplexers are used as one method of reducing the number of logic gates required in a circuit or when a single data line is required to carry two or more different digital signals.

Fig. 8.40 An example of a multiplexer configuration

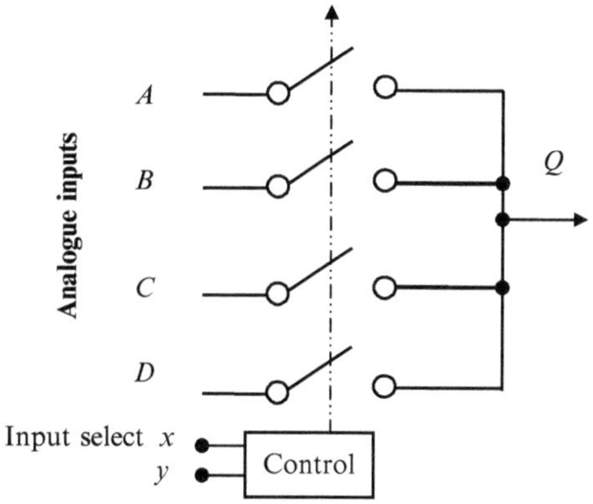

Table 8.18 Truth table of a multiplexer

Addressing		
x	*y*	Input selected
0	0	***A***
1	0	***B***
0	1	***C***
1	1	***D***

Digital Multiplexers are constructed from individual analogue switches encased in a single IC package as opposed to the "mechanical" type selectors such as normal conventional switches and relays. Generally, multiplexers have an even number of data inputs, usually an even power of two, n^2, a number of "control" inputs that correspond with the number of data inputs and according to the binary condition of these control inputs, the appropriate data input is connected directly to the output. An example of a Multiplexer configuration is shown in Fig. 8.40 (Table 8.18).

The Boolean expression for this 4-to-1 Multiplexer above with inputs A to D and data select lines *x*, *y* is given as:

$$Q = A{\cdot}\overline{x}{\cdot}\overline{y} + B{\cdot}x{\cdot}\overline{y} + C{\cdot}\overline{x}{\cdot}y + D{\cdot}x{\cdot}y.$$

In this example at any one instant in time only ONE of the four analogue switches is closed, connecting only one of the input lines *A* to *D* to the single output at *Q*. Which switch is closed depends upon the addressing input code on lines *x* and *y*, so for this example to select input B to the output at *Q*, the binary input address would need to be $x=$logic "**1**" and y=logic "**0**". Adding more control address lines will allow the multiplexer to control more inputs but each control line configuration will connect only **one** input to the output.

Then the implementation of this Boolean expression above using individual logic gates would require the use of seven individual gates consisting of AND, OR and NOT gates, Fig. 8.41.

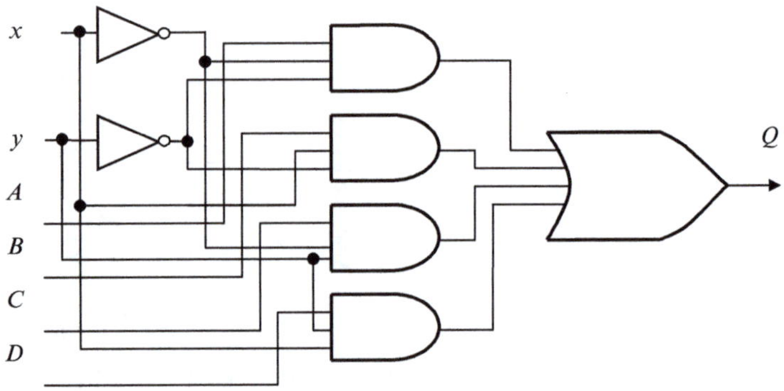

Fig. 8.41 An example of a multiplexer configuration with use of seven individual gates consisting of AND, OR and NOT gates

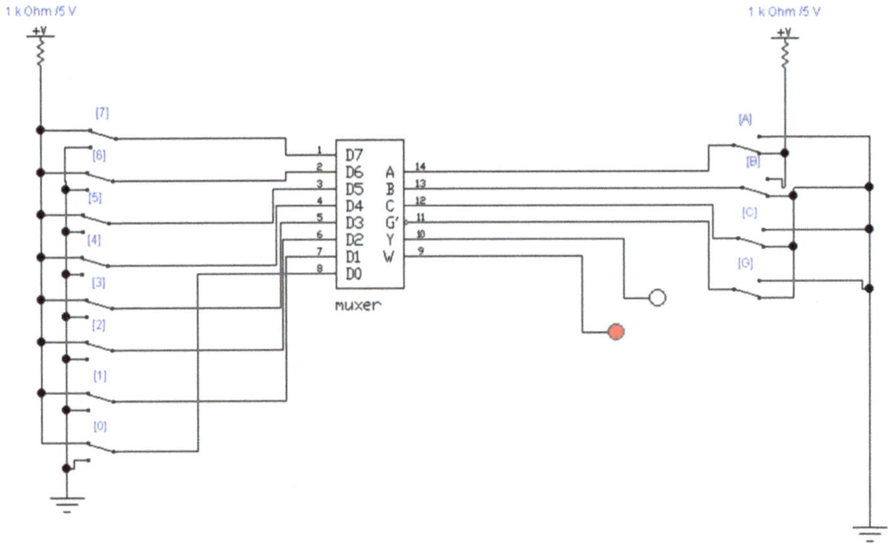

Fig. 8.42 The scheme of the multiplexer (Electronic Workbench 5.12)

Figure 8.42 shows the scheme of the multiplexer. Changing the position of key *A*, *B*, *C*, *G* you can track the change in function at the output of the device – *Y*,*W*.

8.6.2 *The Demultiplexer*

The data distributor, known more commonly as a Demultiplexer or "**Demux**", is the exact opposite of the Multiplexer. The demultiplexer takes one single input data line and then switches it to any one of a number of individual output lines one at a time.

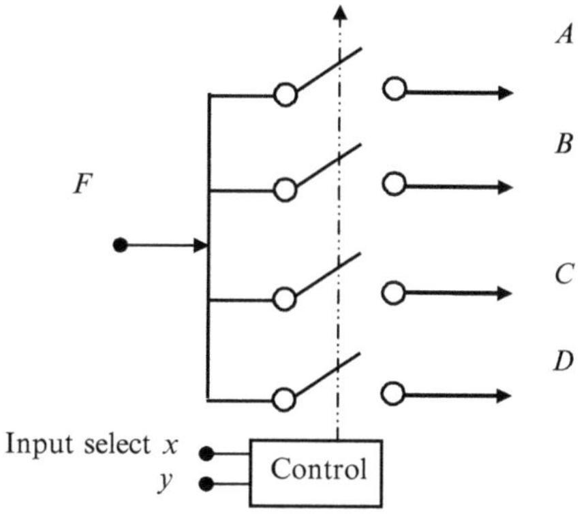

Fig. 8.43 The scheme of the demultiplexer

Table 8.19 Truth table of a demultiplexer

Addressing		Output selected
x	y	
0	0	**A**
1	0	**B**
0	1	**C**
1	1	**D**

The demultiplexer converts a serial data signal at the input to a parallel data at its output lines as shown in Fig. 8.43 (Table 8.19).

$$F = A\cdot\overline{x}\cdot\overline{y} + B\cdot x\cdot\overline{y} + C\cdot\overline{x}\cdot y + D\cdot x\cdot y.$$

The function of the Demultiplexer is to switch one common data input line to any one of the 4 output data lines *A* to *D* in our example above. As with the multiplexer the individual solid state switches are selected by the binary input address code on the output select pins *x* and *y* and by adding more address line inputs it is possible to switch more outputs giving a 1-to-2^n data line output. Some standard demultiplexer IC's also have an "enable output" input pin which disables or prevents the input from being passed to the selected output. Also some have latches built into their outputs to maintain the output logic level after the address inputs have been changed. However, in standard decoder type circuits the address input will determine which single data output will have the same value as the data input with all other data outputs having the value of logic "**0**".

Unlike multiplexers which convert data from a single data line to multiple lines and demultiplexers which convert multiple lines to a single data line, there are devices available which convert data to and from multiple lines.

The implementation of the Boolean expression above using individual logic gates would require the use of six individual gates consisting of AND and NOT gates, Fig. 8.44.

Figure 8.45 shows the scheme of the demultiplexer. Changing the position of key *A*, *B*, *C*, *G* you can track the change in function at the output of the device – **0**–**7**.

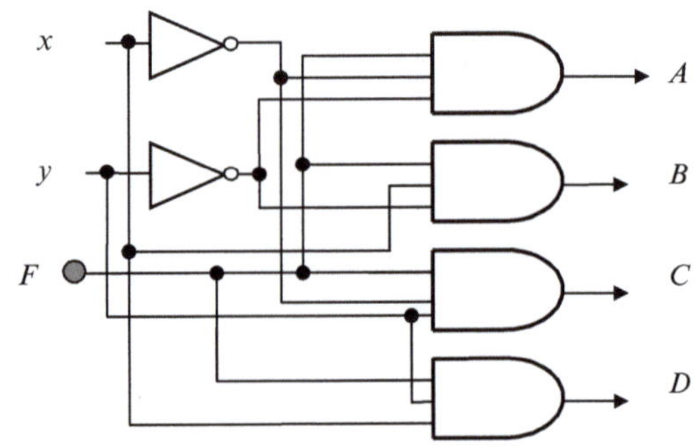

Fig. 8.44 Demultiplexer with using AND gates

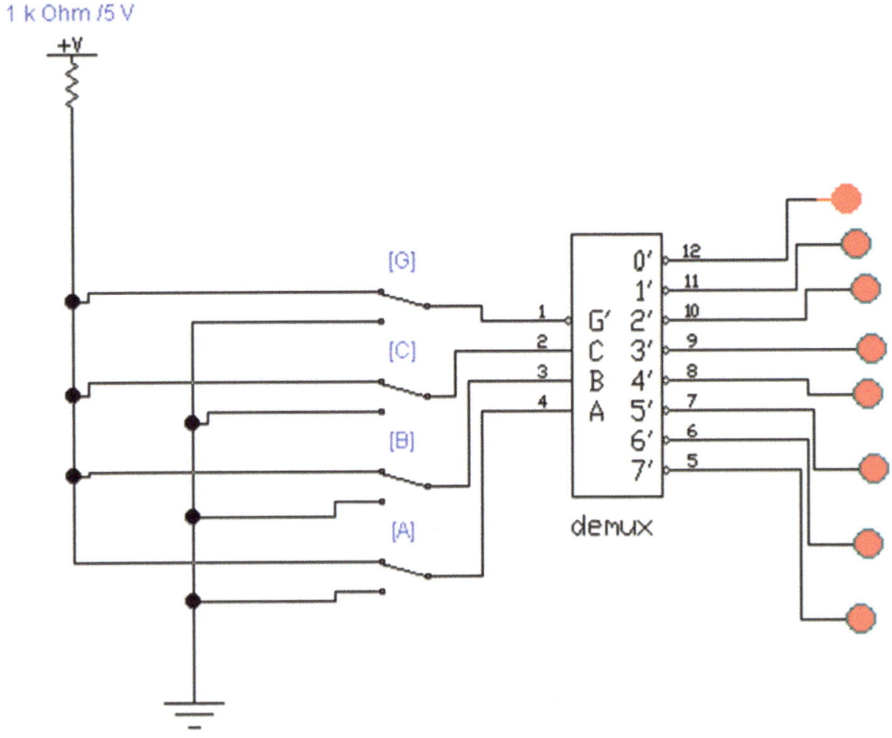

Fig. 8.45 The scheme of the demultiplexer

8.7 Digital Encoder and Decoder

8.7.1 The Digital Encoder

Unlike a multiplexer that selects one individual data input line and then sends that data to a single output line or switch, a Digital Encoder more commonly called a Binary Encoder takes ALL its data inputs one at a time and then converts

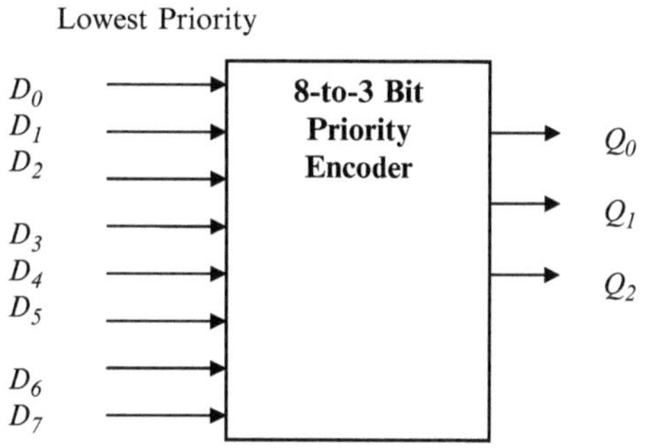

Fig. 8.46 8-to-3 Bit Priority Encoder

them into a single encoded output. So we can say that a binary encoder, is a multi-input combinational logic circuit that converts the logic level "1" data at its inputs into an equivalent binary code at its output. Generally, digital encoders produce outputs of 2-bit, 3-bit or 4-bit codes depending upon the number of data input lines. An "n-bit" binary encoder has 2^n input lines and n-bit output lines with common types that include 4-to-2, 8-to-3 and 16-to-4 line configurations. The output lines of a digital encoder generate the binary equivalent of the input line whose value is equal to "1" and are available to encode either a decimal or hexadecimal input pattern to typically a binary or BCD output code.

One of the main disadvantages of standard digital encoders is that they can generate the wrong output code when there is more than one input present at logic level "**1**".

One simple way to overcome this problem is to "**Prioritise**" the level of each input pin and if there was more than one input at logic level "1" the actual output code would only correspond to the input with the highest designated priority. Then this type of digital encoder is known commonly as a Priority Encoder or P-encoder for short, Fig. 8.46.

Priority Encoders solve the problem mentioned above by allocating a priority level to each input. The encoder output corresponds to the currently active input with the highest priority. So when an input with a higher priority is present, all other inputs with a lower priority will be ignored. Priority encoders come in many forms with an example of an 8-input priority encoder along with its truth Table 8.20, shown below.

Priority encoders are available in standard IC form. Priority encoders output the highest order input first; for example, if input lines "D_2", "D_3" and "D_5" are applied simultaneously the output code would be for input "D_5" ("101") as this has the highest order out of the three inputs. Once input "D_5" had been removed the next highest output code would be for input "D_3" ("011"), and so on.

The Boolean expression for this 8-to-3 encoder above with inputs D_0 to D_7 and outputs Q_0, Q_1, Q_2 is given as:

$$Q_0 = D_1 + D_3 + D_5 + D_7;$$
$$Q_1 = D_2 + D_3 + D_6 + D_7;$$
$$Q_2 = D_4 + D_5 + D_6 + D_7.$$

Table 8.20 Truth table of a priority encoder

Inputs								Outputs		
D_7	D_6	D_5	D_4	D_3	D_2	D_1	D_0	Q_2	Q_1	Q_0
0	0	0	0	0	0	0	1	0	0	0
0	0	0	0	0	0	1	*	0	0	1
0	0	0	0	0	1	*	*	0	1	0
0	0	0	0	1	*	*	*	0	1	1
0	0	0	1	*	*	*	*	1	0	0
0	0	1	*	*	*	*	*	1	0	1
0	1	*	*	*	*	*	*	1	1	0
1	0	*	*	*	*	*	*	1	1	1

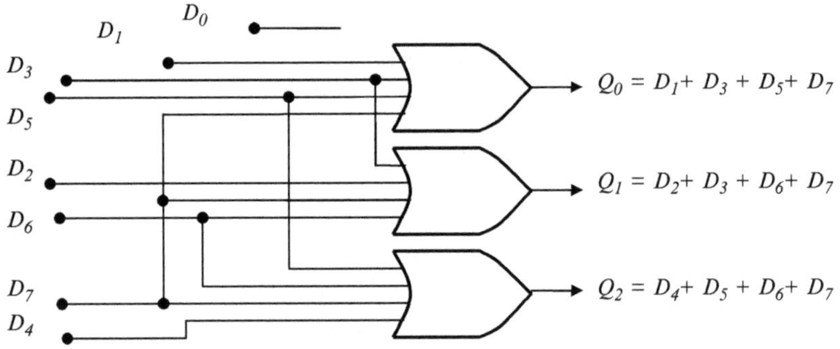

Fig. 8.47 8-to-3 Bit Priority Encoder with using OR gates

Then the implementation of these Boolean expression outputs above using individual OR gates is as follows, Fig. 8.47.

8.7.2 Decoder

A decoder is basically a combinational type logic circuit that converts the binary code data at its input into one of a number of different output lines, one at a time producing an equivalent decimal code at its output. Binary decoders have inputs of 2-bit, 3-bit or 4-bit codes depending upon the number of data input lines, and an n-bit decoder has 2n output lines. Therefore, if it receives n inputs (usually grouped as a binary or Boolean number) it activates one and only one of its 2n outputs based on that input with all other outputs deactivated. A decoder's output code normally has more bits than its input code and practical binary decoder circuits include, 2-to-4, 3-to-8 and 4-to-16 line configurations.

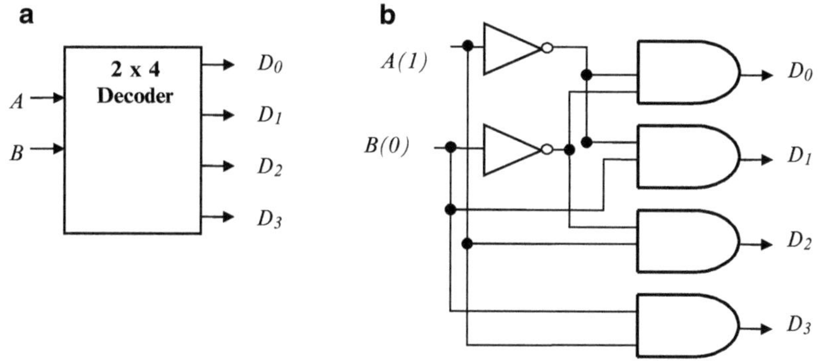

Fig. 8.48 Binary decoder

Table 8.21 Truth table of a 2-to-4 decoder

Binary input		Decoded output			
A	B	D_0	D	D_2	D_3
0	0	1	0	0	0
0	1	0	1	0	0
1	0	0	0	1	0
1	1	0	0	0	1

8.7.2.1 Binary Decoder

A binary decoder converts coded inputs into coded outputs, where the input and output codes are different and decoders are available to "decode" either a Binary or BCD input pattern to typically a Decimal output code. Commonly available BCD-to-Decimal decoders include the TTL 7442 or the CMOS 4028. An example of a 2-to-4 line decoder along with its truth table is given below, Fig. 8.48, Table 8.21. It consists of an array of four NAND gates, one of which is selected for each combination of the input signals A and B.

8.8 Digital Comparator

Another common and very useful combinational logic circuit is that of the Digital Comparator circuit. Digital or Binary Comparators are made up from standard AND, NOR and NOT gates that compare the digital signals present at their input terminals and produce an output depending upon the condition of those inputs. For example, along with being able to add and subtract binary numbers we need to be able to compare them and determine whether the value of input A is greater than, smaller than or equal to the value at input B etc.. The digital comparator accomplishes this

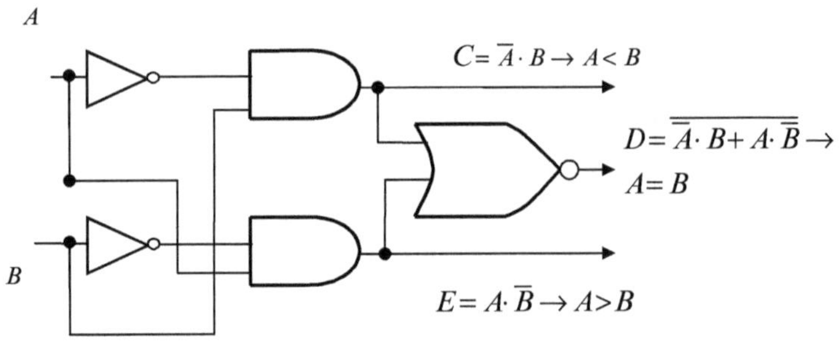

Fig. 8.49 The simple 1-bit comparator

Table 8.22 Truth table of a 1-bit digital comparator

Inputs		Outputs		
A	*B*	*E*	*D*	*C*
0	0	0	1	0
1	0	1	0	0
0	1	0	0	1
1	1	0	1	0

using several logic gates that operate on the principles of Boolean algebra. There are two main types of digital comparator available and these are.

Identity Comparator – is a digital comparator that has only one output terminal for when $A=B$ either "HIGH" $A=B=1$ or "LOW" $A=B=0$.

Magnitude Comparator – is a type of digital comparator that has three output terminals, one each for equality, $A=B$ greater than, $A>B$ and less than $A<B$.

The purpose of a digital comparator is to compare a set of variables or unknown numbers, for example A ($A_1, A_2, A_3, \ldots, A_n$, etc.) against that of a constant or unknown value such as B (B_1, B_2, $B_3, \ldots$, B_n, etc.) and produce an output condition or flag depending upon the result of the comparison. For example, a magnitude comparator of two 1-bits, (A and B) inputs would produce the following three output conditions when compared to each other.

$$A > B, \quad A = B, \quad A < B.$$

This is useful if we want to compare two variables and want to produce an output when any of the above three conditions are achieved. For example, produce an output from a counter when a certain count number is reached. Consider the simple 1-bit comparator below, Fig. 8.49.

Then the operation of a 1-bit digital comparator is given in the following truth table, Table 8.22.

You may notice two distinct features about the comparator from the above truth table. Firstly, the circuit does not distinguish between either two "**0**"s or two "**1**"s as an output; $A=B$ is produced when they are both equal, either $A=B=$"**0**" or $A=B=$"**1**". Secondly, the output condition for $A=B$ resembles that of a commonly available logic gate, the Exclusive-NOR or Ex-NOR function (equivalence) on each of the n-bits giving: $\overline{A \oplus B}$.

Digital comparators actually use Exclusive-NOR gates within their design for comparing their respective pairs of bits. When we are comparing two binary or BCD values or variables against each other, we are comparing the "magnitude" of these values, a logic "**0**" against a logic "1" which is where the term Magnitude Comparator comes from.

As well as comparing individual bits, we can design larger bit comparators by cascading together n of these and produce an n-bit comparator just as we did for the n-bit adder in the previous tutorial. Multi-bit comparators can be constructed to compare whole binary or BCD words to produce an output if one word is larger than, equal to or less than the other. A very good example of this is the 4-bit Magnitude Comparator.

Digital Comparators are used widely in Analogue-to-Digital converters, (ADC) and Arithmetic Logic Units, (ALU) to perform a variety of arithmetic operations.

Appendix A contains examples of programs (Delphi, C#) for simulating the operation of digital automata, presented in this chapter.

Chapter 9
Machines Moore and Mealy

Abstract This chapter presents two finite-state machines (FMS): a Moore FMS (in which output values are determined solely by its current state) and a Mealy FMS (whose output values are determined both by its current state and by the values of its inputs). The chapter contains many examples of designing digital electronic systems, which are at the same time a restricted form of Moore machine (where the state changes only when the global clock signal changes), like a Moore FSM that performs a multiplication or binary division. Additionally the chapter explains the methods of hardware implementation of the constructed Moore FSMs.

In the theory of computation, a Moore machine is a finite-state machine whose output values are determined solely by its current state, Fig. 9.1 This is in contrast to a Mealy machine, whose output values are determined both by its current state and by the values of its inputs, Fig. 9.2.

Where $\alpha_1,\ldots,\alpha_n$ – outputs of memory elements;

$\beta_1,\ldots,\beta_j$ – Boolean functions of excitation of the memory elements;

$W_1,\ldots,W_J$ – output channels of automaton;

j – number of output channels of automaton;

$Z_1,\ldots,Z_m$ – input channels automaton.

9.1 Synthesis of Moore Automata from Graph-Scheme

Since it is impossible to implement machines that have infinite storage capabilities, we shall concentrate on those machines whose past histories can affect their future behavior in only a finite number of ways. We shall study machines that can distinguish among a finite number of classes of input histories and shall refer to these classes as the internal states of the machine. Every finite-state machine, therefore, contains a finite number of memory devices, which store the information regarding the past input history.

B. Borowik et al., *Theory of Digital Automata*, Intelligent Systems, Control and Automation: Science and Engineering 63, DOI 10.1007/978-94-007-5228-3_9,

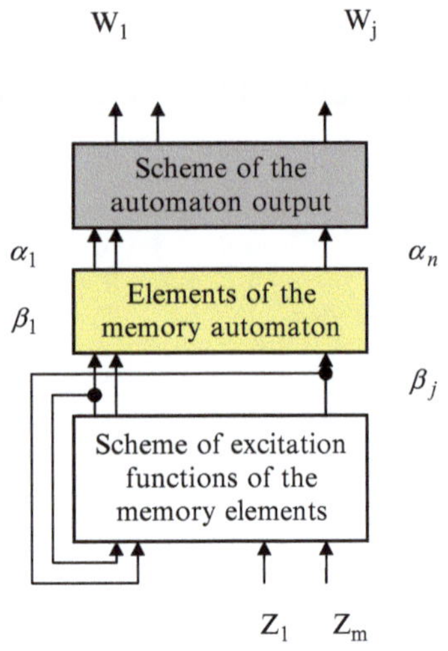

Fig. 9.1 Structural diagram of the Moore machine

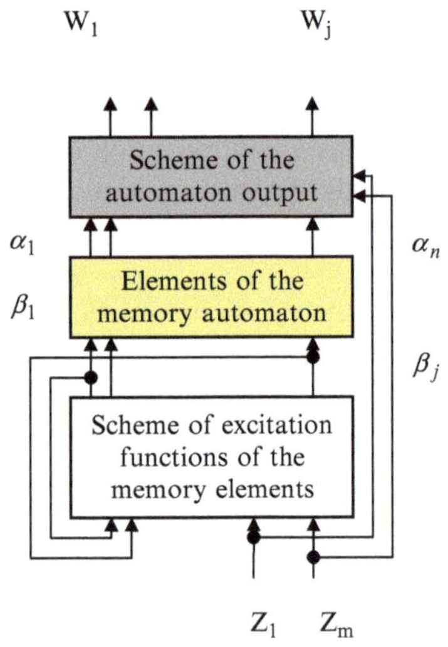

Fig. 9.2 Structural diagram of the Mealy machine

Note that, although we are restricting our attention to machines that have finite storage capacity, no bound has been set on the duration for which a particular input value may affect the future behavior of the machine.

In the theory of computation, a Moore machine is a finite-state machine whose output values are determined solely by its current state, figure. This is in contrast to a Mealy machine, whose output values are determined both by its current state and by the values of its inputs, Figs. 9.1 and 9.2.

Most digital electronic systems are designed as clocked sequential systems. Clocked sequential systems are a restricted form of Moore machine where the state changes only when the global clock signal changes. Typically the current state is stored in flip-flops, and a global clock signal is connected to the "clock" input of the flip-flops. Clocked sequential systems are one way to solve metastability problems. A typical electronic

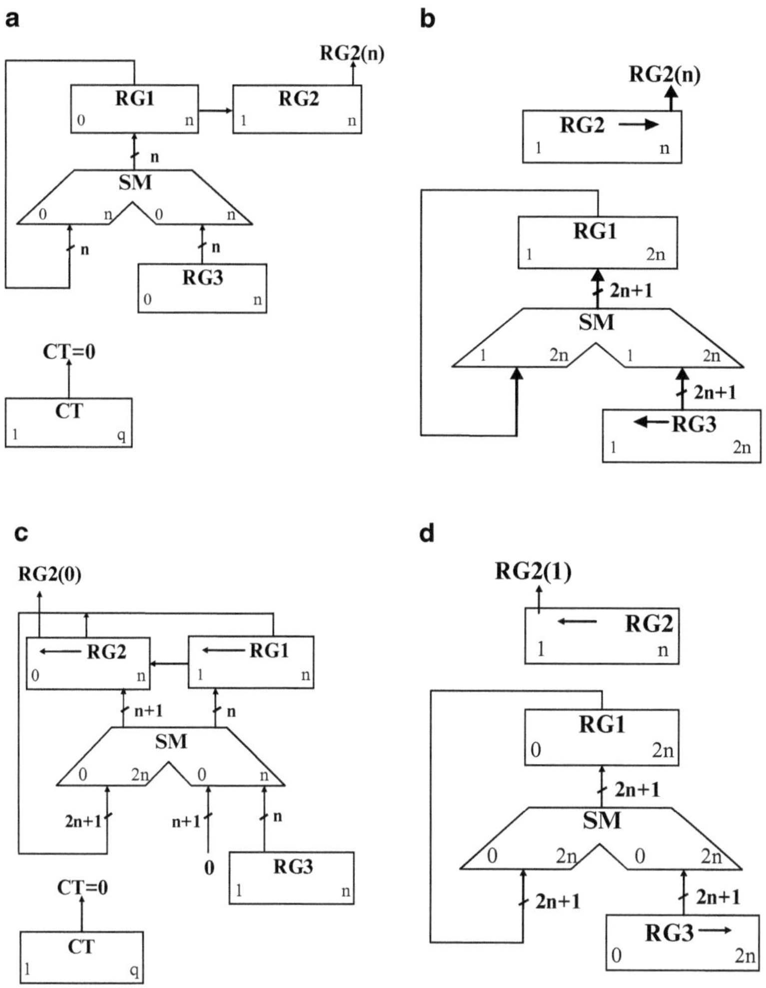

Fig. 9.3 Methods of multiplication. (**a**) 1st method; (**b**) 2nd method; (**c**) 3rd method; (**d**) 4th method

Moore machine includes a combinational logic chain to decode the current state into the outputs. The instant the current state changes, those changes ripple through that chain, and almost instantaneously the outputs change (or don't change). There are design techniques to ensure that no glitches occur on the outputs during that brief period while those changes are rippling through the chain, but most systems are designed so that glitches during that brief transition time are ignored or are irrelevant. The outputs then stay the same indefinitely, until the Moore machine changes state again.

The principle of designing devices to implement the various methods of multiplication are shown in Fig. 9.3

We consider the example of designing devices for multiplying numbers. During the multiplication of numbers in binary code the sign bit and data bits are handled separately. We assume that Y and X – correct binary fraction –

$$X = 0, x_1, x_2, \ldots, x_n,$$

$$Y = 0, y_1, y_2, \ldots, y_n,$$

where x_i, $y_i \in \{0,1\}$.

You can use one of the four methods of multiplication.

1st method

$$Z = Y \cdot X = \Big(\Big(\ldots\Big(\Big(0 + Yx_n\Big)2^{-1} + Yx_{n-1}\Big)2^{-1} + \cdots + Yx_i\Big)2^{-1} + \cdots + Yx_1\Big)2^{-1}$$

2nd method

$$Z = \Big(\Big(\cdots\Big(0 + Y2^{-n}\left(X_n\right)\Big) + Y2^{-n+1}x_{n-1}\Big)\Big) + \cdots + Y2^{-1}x_1$$

3rd method

$$Z = \Big(\cdots\Big(\Big(0 + Y2^{-n}\left(X_1\right)\Big)2 + Y2^{-n}x_2\Big)2 + \cdots + Y2^{-n}x_i\Big)2 + \cdots + Y2^{-n}x_n$$

4th method

$$Z = \Big(\cdots\Big(\Big(0 + Y2^{-1}x_1\Big) + Y2^{-2}x_2\Big) + \cdots + Y2^{-i}x_i\Big) + \cdots + Y2^{-n}x_n.$$

Methods of multiplication on serial binary machines vary from the standard pencil-and-paper method to multiplication by a fast multiplier. The time taken for a multiplication by both these methods is independent of the arrangement of digits in the multiplier. For some methods, however, the time is not invariant, so that for an optimum-coded machine using a delay-type store it is essential to know the expected time for a multiplication, so that maximum time-saving may be obtained, by inserting the next order in the optimum position. In a machine of this type each order includes an indication of the operation to be performed and the address of the next order; other addresses may be specified also. For the majority of operations, e.g. addition, subtraction, doubling, halving, some logical orders and magnitude tests, the time taken is known and fixed for the same operation. Thus the next order may be placed, subject to availability, so that it may be called in immediately the previous operation is completed. Further, if the next order is placed in an earlier position, it will not be available until after a complete period of the delay store, and if in a later position, a certain amount of time will be wasted. For operations such as multiplication and division, with variable time for completion, the optimum position for the next order is such that over all possible times the expected time is minimized. It may be necessary in the following methods to allow time for control instructions, but these are functions of the overall construction of the computer and, as such, do not affect the basic time for a multiplication.

To see how this can be done, notice that the result of multiplying two n-digit positive binary numbers, the result may be as long as 2n digit long. Let us denote the first factor by x, the second by y, and the result by Z.

When we implement multiplication in digital computers we have ways to change the method easily. Firstly, instead of providing registers to store and add intermediators generated in between the multiplication, we can have adder to add the intermediators simultaneously and store them then in registers. This saves memory. Secondly, instead of shifting the multiplicand to the left it is suggested to shift the partial product to the right. It makes the relative positions for the partial product and multiplicand. Thirdly, when the corresponding bit of the multiplier is 0, then there is no need to multiply the number as the result doesn't have any difference form it.

Principles of designing devices that implement the various methods of multiplication are shown in Fig. 8.28.

Where RG1 – register; RG2 – register X; RG3 – register Y.

These registers along with two other registers make up the complete implementation. Principles of designing devices that implement the various methods of multiplication are shown in Fig. 8.28 where

RG1 – register $Z = Y \cdot X$;
RG2 – register X;
RG3 – register Y.

Example 1. Build a circuit that simulates activity of a Moore automaton given by the following graph-scheme, Fig. 9.4.
Step 1: Assign the marks to the scheme:

- Mark the node Start and End by symbol – a_1;
- Mark the operator nodes by symbols $a_2, \ldots, a_m$, each by one symbol.

Step 2: Derive the transitions and outputs tables. The transition functions are formed as conjunctions of the ways from a_m to a_s. Output data: Graph of Moore automata, Fig. 9.5.
Step 3: We obtain a transitions/outputs table (Tables 9.1 and 9.2):

Example 2. We consider the example synthesis of the Moore automata for multiplication of two binary numbers.

We introduce the following notation:
SM – adder;
CT – counter;
RG, RG1 – registers;
n – number of cycles.
Let $A = 1101_2$, $B = 101_2$. Then

			1	1	0	1_2	Multiplier
				1	0	1_2	Multiplicand
			1	1	0	1	Partial-products
		0	0	0	0		
	1	1	0	1			
1	0	0	0	0	0	1_2	**Product**

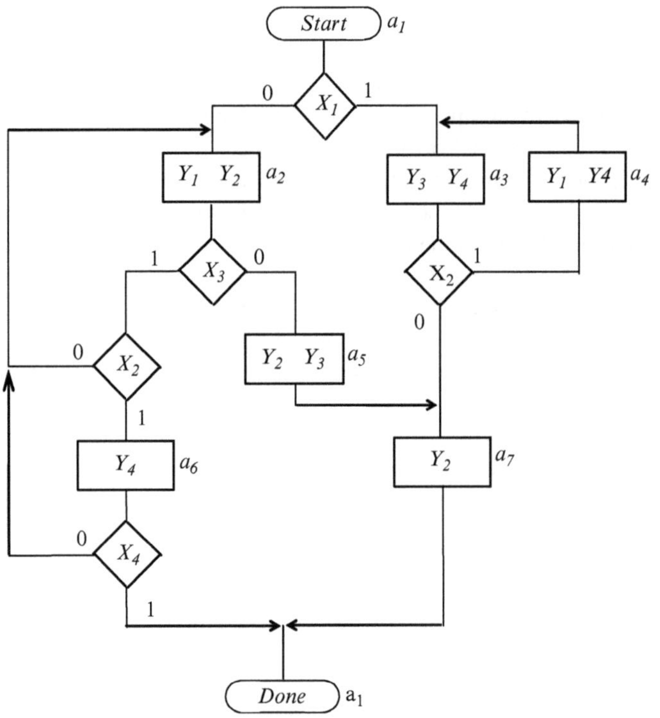

Fig. 9.4 Graph-scheme of a Moore automaton

Fig. 9.5 Graph of a Moore automaton

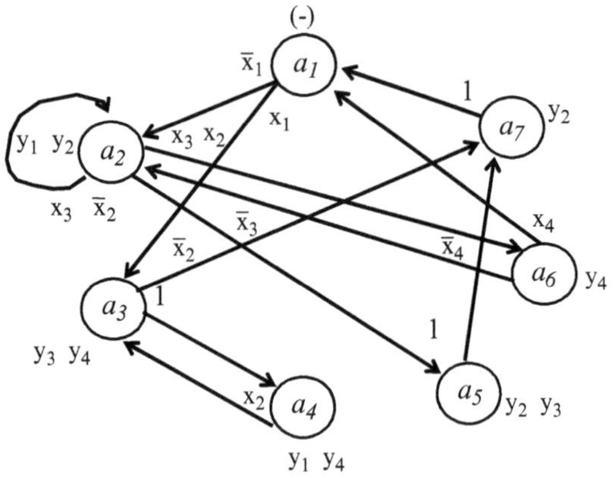

Table 9.1 Moore automaton state transition table (direct transition table)

$a_m(Y)$	a_s	x
$a_1(--)$	a_2	$\bar{x}_1$
	a_3	$\bar{x}_1$
$a_2(y_1y_2)$	a_2	$x_3 \cdot \bar{x}_2$
	a_5	$\bar{x}_3$
	a_6	$\bar{x}_3 \cdot x_2$
$a_3(y_3y_4)$	a_4	x_2
	a_7	$\bar{x}_2$
$a_4(y_1y_4)$	a_3	1
$a_5(y_2y_3)$	a_7	1
$a_6(y_4)$	a_1	x_4
	a_2	$\bar{x}_4$
$a_7(y_2)$	a_1	1

Table 9.2 Moore automaton state transition table (reverse transition table)

a_m	$a_s(y)$	x
a_6	$a_1(-)$	x_4
a_7		1
a_1	$a2(y_1y_2)$	$\bar{x}_1$
a_2		$x_3 \cdot \bar{x}_2$
a_6		$\bar{x}_4$
a_1	$a_3(y_3y_4)$	x_1
a_4		1
a_3	$a_4(y_1y_4)$	x_2
a_2	$a_5(y_2y_3)$	$\bar{x}_3$
a_2	$a_6(y_4)$	$x_3 \cdot x_2$
a_3	$a_7(y_2)$	$\bar{x}_2$
a_5		1

The Decision considered in the above example is in this instance realized on the following scheme (Figs. 9.6, 9.7, 9.8):

RG2	*SM* (adder)	*RG1* (LSB)	Comment
1101	00000	101	
	± 1101		
	01101		Addition
		110	Shift (*RG1*) and (*SM*)
	00110		
	00011	011	Shift (*RG1*) and (*SM*)
	± 1101		Addition
	10000		
	01000	001	Shift (*RG1*) and (*SM*)
	1000	001	**Product**

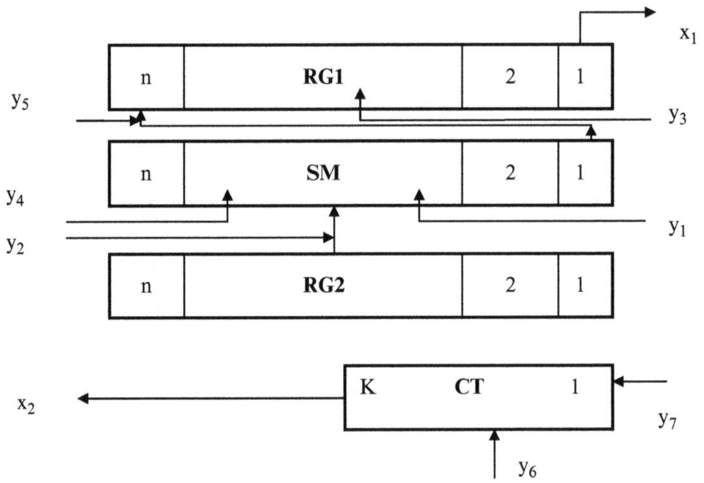

Fig. 9.6 The block diagram of an operational device that implements the multiplication

Fig. 9.7 Graph-scheme of multiplying two numbers

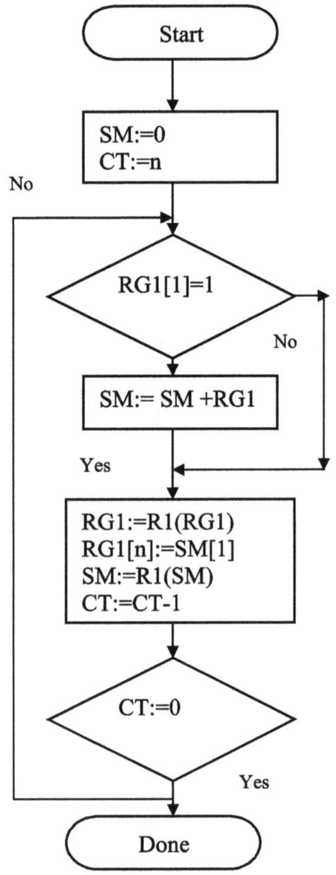

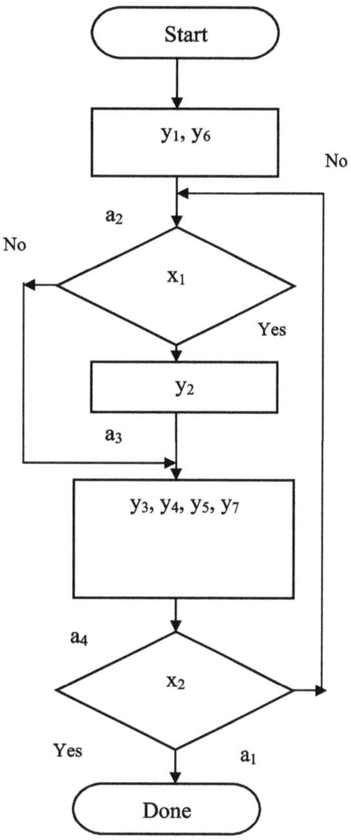

Fig. 9.8 Graph-scheme of coded multiplication algorithm of two numbers

Where

y_1 – installation of *SM* in the zero state (*SM*: = 0);
y_2 – *SM*: = *SM*+*RG*2;
y_3 – transfer the contents of the first register in the first register with a shift to the right (*RG*1=*R*1(*RG*1));
y_4 – shift the contents of the adder to the right (*SM*: = *R*1 (*SM*));
y_5 – shift LSB to MSB adder register (*RG*1[*n*]:= *SM* [1]);
y_6 – input to the counter number *n*;
y_7 – signal operations account (*CT*: = *CT*–1).

Based on the trigger **R – S** of the transition table, fill columns **R and S**, Table 9.3 (Figs. 9.9 and 9.10).

Also $Y_1=y_1, y_6$; $Y_2=y_2$; $Y_3=y_3, y_4, y_5, y_7$.

The project requires an algorithm capable of comparing two circuits. It may need to search thousands of circuits, so it must be as efficient as possible. Furthermore, it must correctly find any sort of analogue circuit, not merely all of those with particular

Table 9.3 A finite-state machine (FSM)

Initial state (a_m)	The code of the initial state (Q_m)		State transition (a_s)	Status code transition $K(a_s)$		Input signal $x\binom{a_m,}{a_s,}$	Output signal $y\binom{a_m,}{a_s,}$	Initialization function trigger $\binom{a_{m,}}{a_s}$	
	Q_2	Q_1		Q_2	Q_1			R	S
a_1	0	0	a_2	0	1	1	y_1	–	S_1
a_2	0	1	a_3	1	0	x_1	y_2	R_2	S_1
a_3	1	0	a_4	1	1	1	y_3	–	S_1
a_4	1	1	a_2	0	1	$\overline{x}_2$	–	R_2	–
a_4	1	1	a_1	0	0	x_2	–	R_1R_2	–
a_2	0	1	a_4	1	1	$\overline{x}_1$	y_3	–	S_2

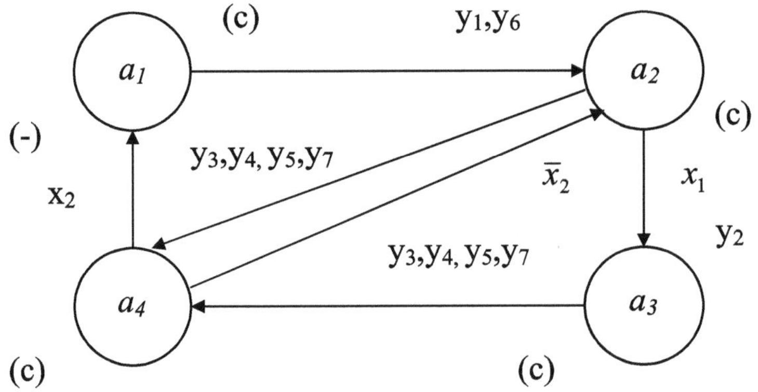

Fig. 9.9 Graph-scheme of coded multiplication algorithm of two numbers

properties, since it is impossible to know every circuit that may be added to the repository in the future, or indeed the circuits that will be searched for.

It is not easy for a computer to determine the function of an analogue circuit. A computer can be given access to every aspect of a circuit that a human would be able to see: component values, interconnections, perhaps even component locations so that the circuit can be drawn on screen.

However, a computer cannot interpret this information as easily as an experienced engineer.

There are some circuits that are easily compared. Digital circuits are a special type of analogue circuit. It is not difficult for a computer to examine a combinatorial digital circuit. A computer can always work out the minimum logical function that such a circuit provides, and compute truth tables. This type of circuit has discrete inputs and outputs, each of which can only take two values.

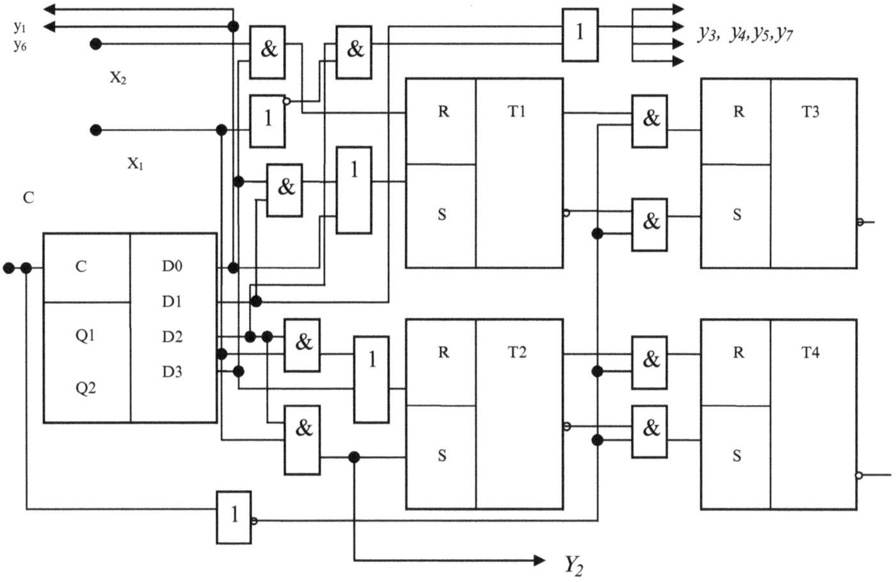

Fig. 9.10 Functional circuit of the multiplication of binary numbers

Table 9.4 Table of state automation and state trigger

State automaton (a_m)	State trigger	
	Q_2	Q_1
a_1	0	0
a_2	0	1
a_3	1	0
a_4	1	1

Combinatorial circuits can thus be compared in terms of the minimal representation of their logical function, or in terms of their truth tables. However, this is not possible for non-combinatorial digital circuits: those with some type of memory or internal state. A logical function or truth table could only be drawn for such circuits if its parameters included all the values of the internal state.

In an analogue circuit, a truth table can never be derived, because all inputs and outputs have real values. Voltage and current are continuous quantities which may take any real-numbered value. Nor is it possible, in general, to reduce an analogue circuit to a mathematical function which could be compared more easily.

An electronic circuit is easily expressed as a graph: an example of one possible representation was illustrated earlier in Figure 9.9. Since this is the case, existing methods for solving subgraph isomorphism problems can be applied to comparing circuits.

Thus, based on data in Tables 9.3 and 9.4 we obtain the following system of equations:

$$\left.\begin{array}{ll} y_1 = \bar{Q}_1 \cdot \bar{Q}_2; & R_1 = Q_1 \cdot Q_2 \cdot x_2; \\ y_2 = Q_1 \cdot \bar{Q}_2 \cdot x_1; & R_2 = Q_1 \cdot Q_2 \cdot \bar{x}_2 \vee Q_1 \cdot Q_2 \cdot x_2 \vee Q_1 \cdot Q_2 \cdot x_1; \\ y_3 = \bar{x}_1 \cdot Q_1 \cdot \bar{Q}_2 \vee \bar{Q}_1 \cdot Q_2; & S_1 = \bar{Q}_1 \cdot \bar{Q}_2 \vee \bar{Q}_1 \cdot Q_2 \vee Q_1 \cdot Q_2 \cdot x_1; \\ & S_1 = \bar{Q}_2 \cdot Q_1 \cdot \bar{x}_1. \end{array}\right\} \quad (9.1)$$

Add to the decoder circuit of automaton states:

$$a_0 = \bar{Q}_1 \cdot \bar{Q}_2; \quad a_1 = \bar{Q}_1 \cdot Q_2; \quad a_2 = Q_1 \cdot \bar{Q}_2; \quad a_3 = Q_1 \cdot Q_2;$$

Thus, the system of Eq. 9.1 can be simplified:

$$\left.\begin{array}{c} R_1 = a_3 \cdot x_2 \cdot C; \\ R_2 = a_3 \cdot \bar{x}_2 \cdot C \vee a_3 \cdot x_2 \cdot C \vee Q_1 \cdot \bar{Q}_2 \cdot x_1 = a_3 \cdot C \vee a_2 \cdot x_1 \cdot C; \\ S_1 = a_0 \cdot C \vee a_1 \cdot C \vee x_1 \cdot a_2 \cdot C; \\ S_2 = a_2 \cdot x_1 \cdot C; \\ y_1 = a_0 \cdot C; \quad y_2 = x_1 \cdot a_2 \cdot C; \quad y_3 = \bar{x}_1 \cdot a_2 \cdot C \vee a_1 \cdot C. \end{array}\right\} \quad (9.2)$$

Division, similar to multiplication we can do division as shown below.

Of all the elemental operations, division is the most complicated and can consume the most resources (in either silicon, to implement the algorithm in hardware, or in time, to implement the algorithm in software). In many computer applications, division is less frequently used than addition, subtraction or multiplication. As a result, some microprocessors that are designed for digital signal processing (DSP) or embedded processor applications do not have a divide instruction (they also usually omit floating point support as well).

We outline the basic algorithmization and programming principles for logic synthesis of the Moore automata for binary division.

Example 3. We outline the basic algorithmization and programming principles for logic synthesis of the Moore automata for binary division.

All algorithms are categorized into two main sections. One is known as slow division while the other one is known as fast division. Both of these algorithms have their own unique working procedure through which they perform all of their tasks. For example, a slow division algorithm always produces only a single digit of each final quotient. Some of the famous examples of a slow division algorithm are restoring and SRT. While on the other hand, a fast division algorithm follows the rule of closest possible approximated value relative to the finally produced quotient and produces as many digits as it can which are in twice pattern of final outcome quotient.

We assume that in an arithmetic division operation involving operands – $C = A/B$.

In addition, the digital device must generate a symptoms result in binary variables:

Fig. 9.11 Binary code

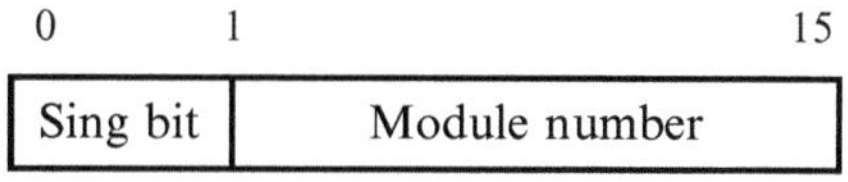

- Z – *a zero result*;
- S – *a negative result*;
- OV – *a sign of overcrowding*.

Algebraic division algorithm operations are designed for 16-bit binary numbers with fixed point (Fig. 9.11).

In computer science, the sign bit is a bit in a computer numbering format that indicates the sign of a number. In IEEE format, the sign bit is the leftmost bit (most significant bit). Typically if the sign bit is 1 the number is negative (in the case of two's complement integers) or non-positive (for ones' complement integers, sign-magnitude integers, and floating point numbers), while 0 indicates a positive number.

Thus, in operations involving the following variables:

- $A = a_0a_1a_2\ldots a_{15}$ – first operand (**dividend**);
- $B = b_0b_1b_2\ldots b_{15}$ – second operand (**divisor**);
- $C = c_0c_1c_2\ldots c_{15}$ – result of the operation division (**Also, there remains a** W);
- $D = d_0d_1d_2\ldots d_{15}$ – variable, which accumulates C;
- a_0, b_0, c_0 – sign bits.

The sign of the result of division can be found from the expression:

$$c_0 = a_0\overline{b}_0 \vee a_0\overline{b}_0.$$

The graph-scheme of the division operation is shown in Fig. 9.12.

For reduction of mean time for a division operation, use a method that does not recover the remainder, an algorithm for which follows.

(1)Define the sign a quotient by summation over module two contents sign category done and divisor.
(2)From done to subtract the divisor. If the remainder $W < 0$, go to point 3. Otherwise calculation to finish.
(3)Remember the sign of the remainder (W).
(4)Shift the remainder (W) on one category to the left.
(5)Assign the divisor a sign, the inverse sign of the remainder, remembered in Step 2.
(6)Pack the shifted remainder and divisor (with provision for sign).
(7)Assign the numeral a quotient importance, opposite code of the sign of the remainder.
(8)Repeat Steps 3–7 until required accuracy of the calculation is achieved.

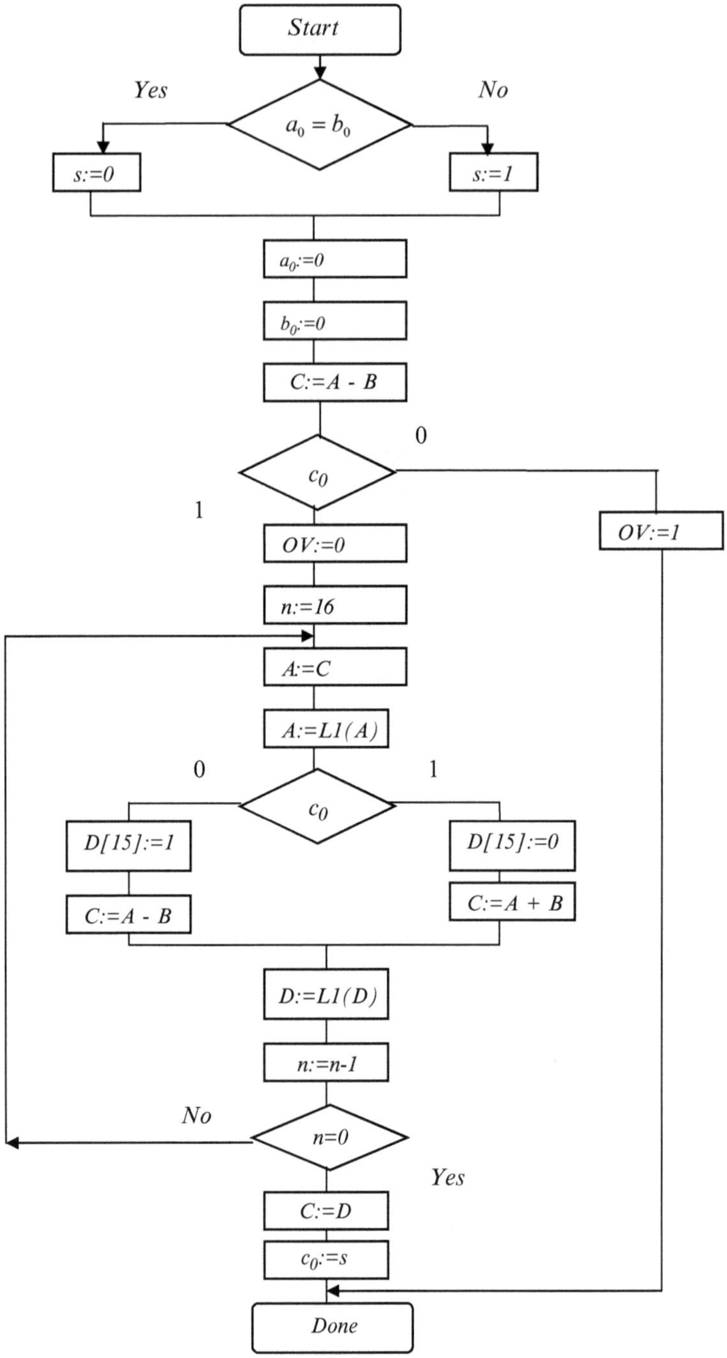

Fig. 9.12 Graph-scheme of the division operation

The Decision considered in the above example is in this instance realized in the following scheme:

$[A]_{tc}$	+00,1001	00,1100
$[-B]_{tc}$ (*subtraction*)	11,0011	00,10110
$W<0$	10	
	proper fraction	
W shift to the left	+11,1000	
Addition $[B]_{tc}$	00,1101	
$W>0$	00,0101	
W shift to the left	+00,1010	
Subtraction	11,0011	
$W<0$		
	11,1101	
W shift to the left	+11,1010	
Addition	00,1101	
$W>0$		
	00,0111	
W shift to the left	+00,1110	
Subtraction	11,0011	
$W>0$		
	00,0001	
W shift to the left	+00,0010	
Subtraction	11,0011	
$W<0$		
	11,0101	

Result $C = A{:}B =$ **0,1011**

Next, determine which sequence of micro operations to be implemented to develop a framework to perform the division is provided by the algorithm of Fig. 9.3. The simplest solution - to keep the topology of the graph algorithm and replace the contents of its operator - the appropriate logical conditions.

Thus obtained graph is called firmware and treated as an input to the design of the receiver (firmware) of the automaton. In this case, the contents of the vertex operator corresponding to the action performed by the device in one step of discrete time. The design of digital systems usually aim to achieve a top speed of their work.

The structure of the automaton should include the following elements:

- Two hexadecimal registers *PгA* and *PгB* for storing the input operands and intermediate results, and the *PгA* register should provide an opportunity to shift its contents to the left;
- Hexadecimal Register *PгC* to accommodate the results of arithmetic operations of addition or subtraction. At the end of the operation it will post the result;

Table 9.5 A complete list of micro-and logical conditions

Micro-operation	Action	Micro-operation	Action	Logical condition	Action
y_1	s:=0	y_{10}	A:=$L1(A)$	x_1	a_0:=b_0
y_2	s:=1	y_{11}	$D[15]$:=1	x_2	c_0
y_3	a_0:=0	y_{12}	$D[15]$:=0	x_3	*Сч n*:=0
y_4	b_0:=$\overline{0}$	y_{13}	C:=$A+B$		
y_5	C:=$R+S$	y_{14}	D:=$L1$(D)		
y_6	OV:=0	y_{15}	*Сч n*:=Сч−1		
y_7	OV:=1	y_{16}	C:=D		
y_8	n:=16	y_{17}	c_0:= s		
y_9	A:=C				

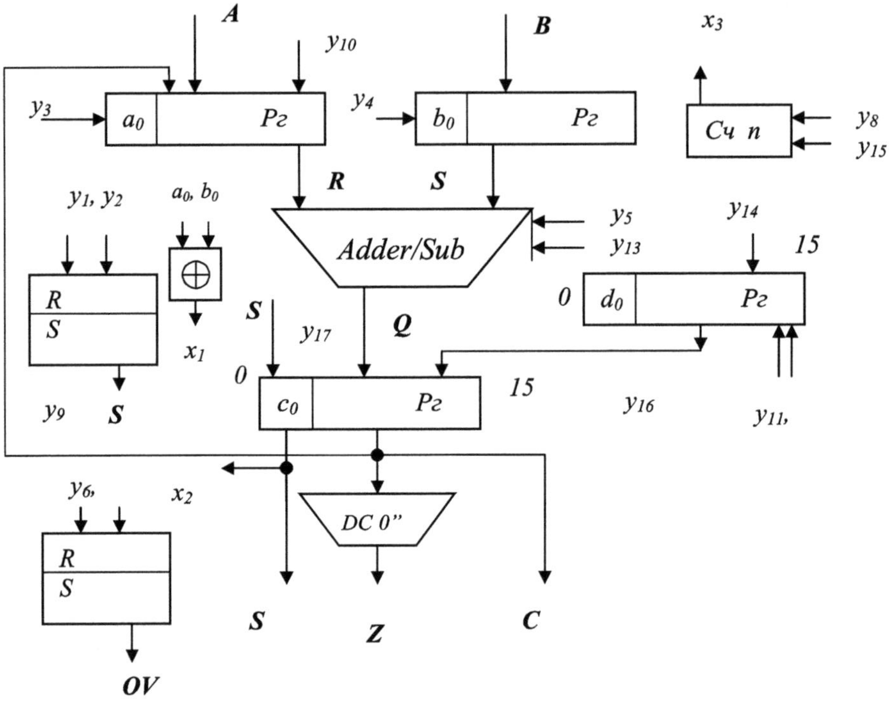

Fig. 9.13 Links between elements of Moore FSM and micro-operations

- Hexadecimal register *PгD* with the ability to left shift code (to accommodate *C*);
- Hexadecimal binary parallel adder/subtractor *Adder/Sub*;
- Four-digit down counter *Сч n*;
- Flip-flop the overflow *Tг OV* to store the overflow trait word length;
- Flip-flop of the sign *Tг s*;
- Scheme comparison "equals" sign bit source operands;
- decoder *DC* "0" zero combination in the ranks *C*[1:15].

Table 9.5 shows a complete list of micro-and logical conditions.

Links between elements of Moore FSM and micro-operations are shown in Fig. 9.13.

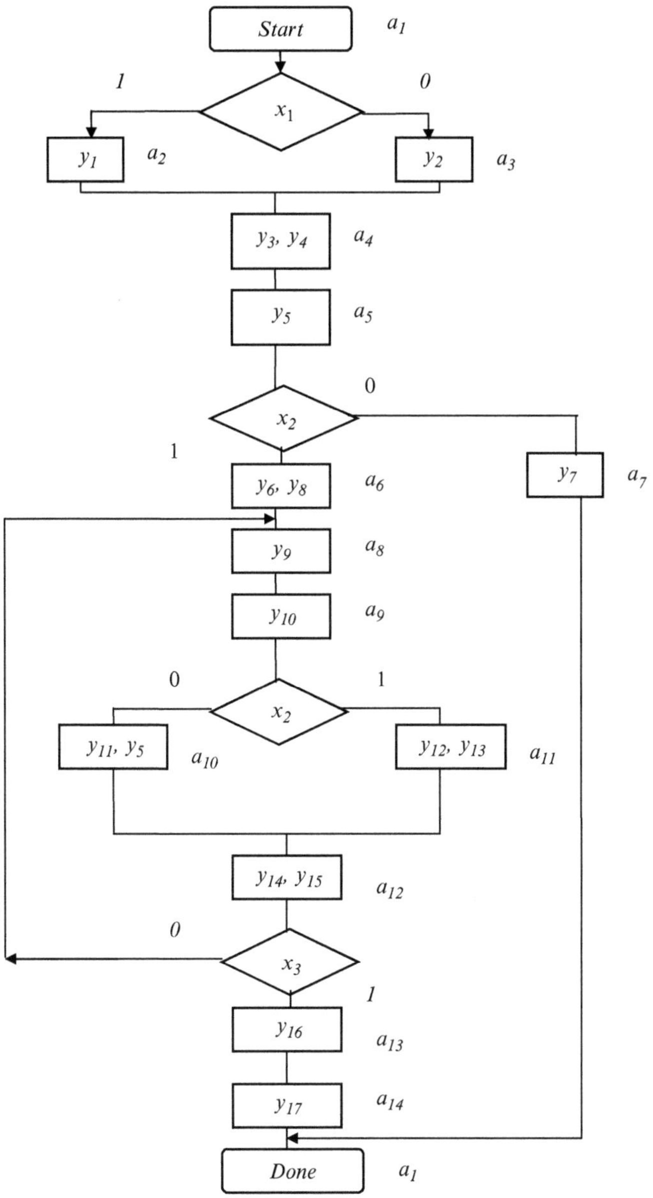

Fig. 9.14 Graph-scheme of a division algorithm

Step 1.

Mark out the microprogram division.

Assign the marks to the scheme, Fig. 9.14.

Step 2.

We construct the graph, we are actually given alphabet of internal states and input symbols and determine the transition function. To set the output symbols of

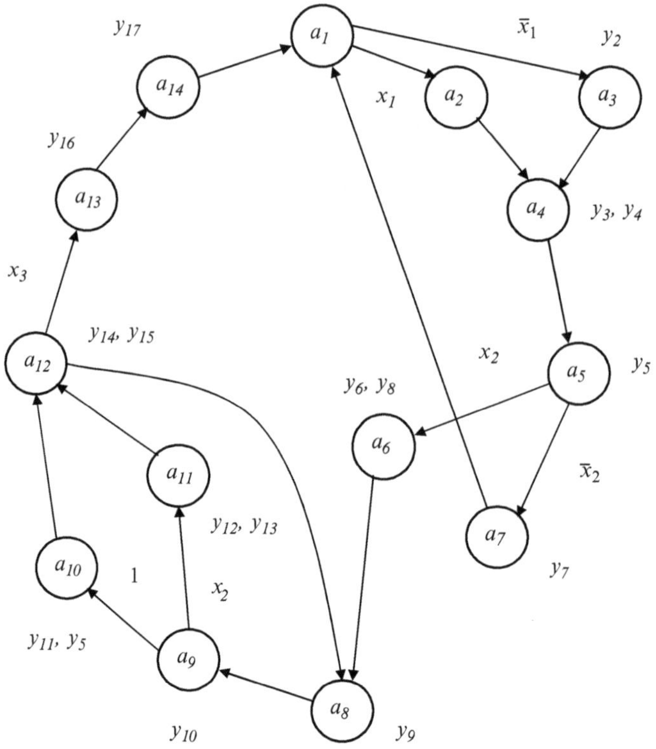

Fig. 9.15 Graph-scheme of division (Moore FSM)

the alphabet and the output function (for Moore FSM output function depends only on the states) should be compared with each vertex in the machine as the output character contents of the corresponding vertex operator firmware. Thus, we obtain the graph firmware machine, which is shown in Fig. 9.15.

A synthesizable state machine may be coded many ways. Two of the most common, easily understood and efficient methods are two-always block and one-always block state machines.

The easiest method to understand and implement is the two-always block state machine with output assignments included in either the combinational next-state always block or separate continuous-assignment outputs.

Step 3.

Perform the coding of states of the digital automata (Table 9.6).

Step 4.

The choice of memory elements. Choosing a D flip-flop. Construct the transition table automaton (Table 9.7).

Step 5.

Development of a combinational circuit.

Table 9.6 Moore automaton state transition table (reverse transition table)

State automaton	Code $T_1\,T_2\,T_3\,T_4$	State automaton	Code $T_1\,T_2\,T_3\,T_4$
a_1	0001	a_8	1000
a_2	0010	a_9	1001
a_3	0011	a_{10}	1010
a_4	0100	a_{11}	1011
a_5	0101	a_{12}	1100
a_6	0110	a_{13}	1101
a_7	0111	a_{14}	1110

Table 9.7 Flip-flop Excitation Tables

The initial state of automaton	Jump condition	State transition	Excitation function of the flip-flop			
			D_1	D_2	D_3	D_4
(a_1) 0001	x_1	(a_2) 0010	0	0	1	0
	$\overline{x}_1$	(a_3) 0011	0	0	1	1
(a_2) 0010	1	(a_4) 0100	0	1	0	0
(a_3) 0011	1	(a_4) 0100	0	1	0	0
(a_4) 0100	1	(a_5) 0101	0	1	0	1
(a_5) 0101	x_2	(a_6) 0110	0	1	1	0
	$\overline{x}_2$	(a_7) 0111	0	1	1	1
(a_6) 0110	1	(a_8) 1000	1	0	0	0
(a_7) 0111	1	(a_1) 0001	0	0	0	1
(a_8) 1000	1	(a_9) 1001	1	0	0	1
(a_9) 1001	x_2	(a_{11}) 1011	1	0	1	1
	$\overline{x}_2$	(a_{10}) 1010	1	0	1	0
(a_{10}) 1010	1	(a_{12}) 1100	1	1	0	0
(a_{11}) 1011	1	(a_{12}) 1100	1	1	0	0
(a_{12}) 1100	x_3	(a_{13}) 1101	1	1	0	1
	$\overline{x}_3$	(a_8) 1000	1	0	0	0
(a_{13}) 1101	1	(a_{14}) 1110	1	1	1	0
(a_{14}) 1110	1	(a_1) 0001	0	0	0	1

Stage of minimization we're missing. You can do it yourself. For example, use Karnaugh Maps.

Excitation function of the flip-flop:

$$D_1 = a_6 \vee a_8 \vee a_9 \vee a_{10} \vee a_{11} \vee a_{12} \vee a_{13}.$$

$$D_2 = a_2 \vee a_3 \vee a_4 \vee a_5 \vee a_{10} \vee a_{11} \vee a_{12}x_3 \vee a_{13}.$$

$$D_3 = a_1 \vee a_5 \vee a_9 \vee a_{13}.$$

$$D_4 = a_1\overline{x}_1 \vee a_4 \vee a_5\overline{x}_2 \vee a_7 \vee a_8 \vee a_9 x_2 \vee a_{12}x_3 \vee a_{14}$$

Output function:

$$
\begin{aligned}
y_1 &= a_2, \\
y_2 &= a_3, \\
y_3 &= a_4, \\
y_4 &= a_4, \\
y_5 &= a_5 \vee a_{10}, \\
y_6 &= a_6, \\
y_7 &= a_7 \\
y_8 &= a_6, \\
y_9 &= a_8, \\
y_{10} &= a_9, \\
y_{11} &= a_{10}, \\
y_{12} &= a_{11}, \\
y_{13} &= a_{11}, \\
y_{14} &= a_{12} \\
y_{15} &= a_{12}, \\
y_{16} &= a_{13}, \\
y_{17} &= a_{14}.
\end{aligned}
$$

Step 6.

Development of a functional diagram of the device (Fig. 9.16).

A function block diagram (FBD) of division algorithm is a block diagram that describes a function between input variables and output variables. A function is described as a set of elementary blocks. Input and output variables are connected to blocks by connection lines. An output of a block may also be connected to an input of another block.

Inputs and outputs of the blocks (logic gates, flip-flops, decoder) are wired together with connection lines, or links. Single lines may be used to connect two logical points of the diagram:

- an input variable and an input of a block;
- an output of a block and an input of another block;
- an output of a block and an output variable.

The connection is oriented, meaning that the line carries associated data from the left end to the right end. The left and right ends of the connection line must be of the same type.

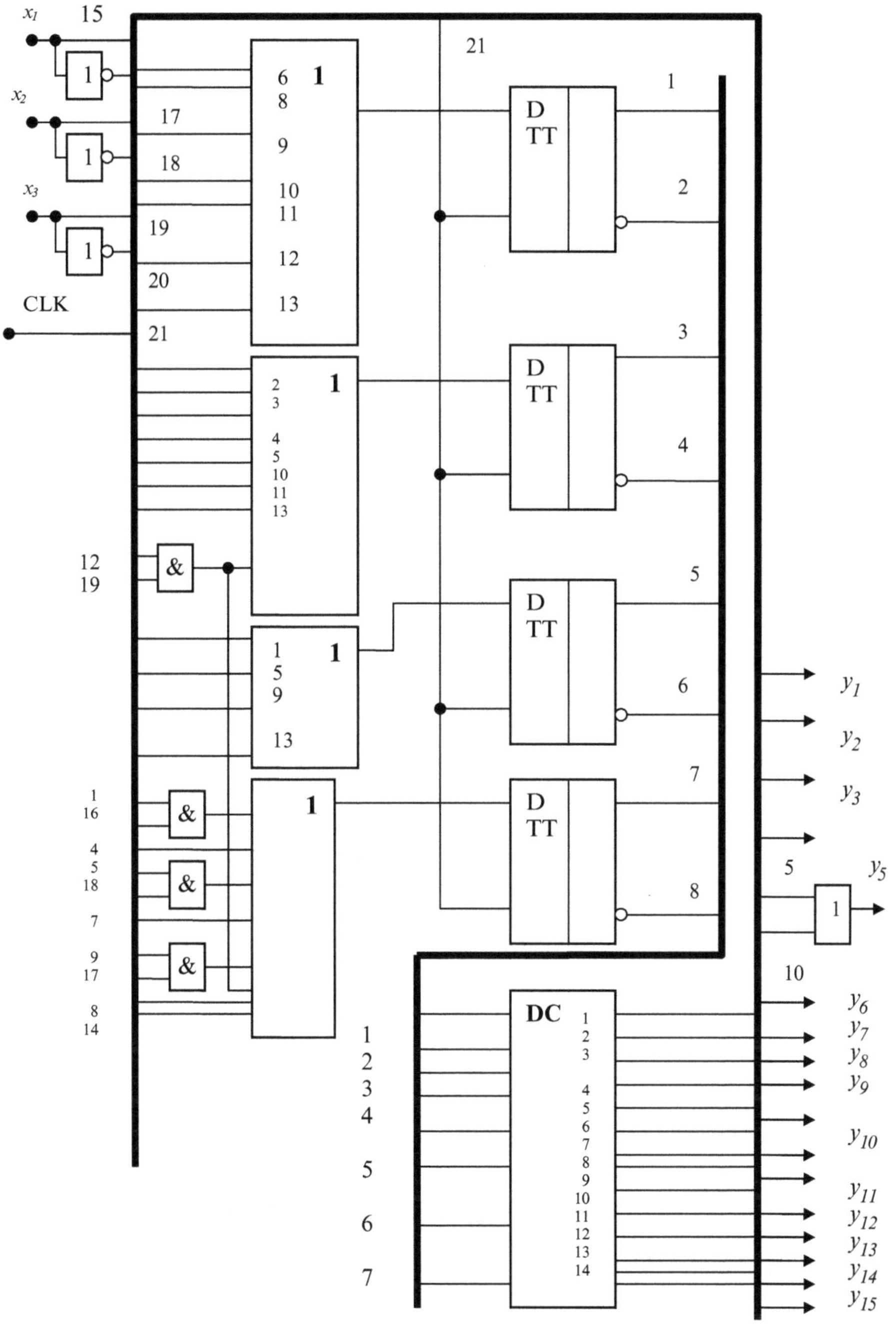

Fig. 9.16 Functional diagram

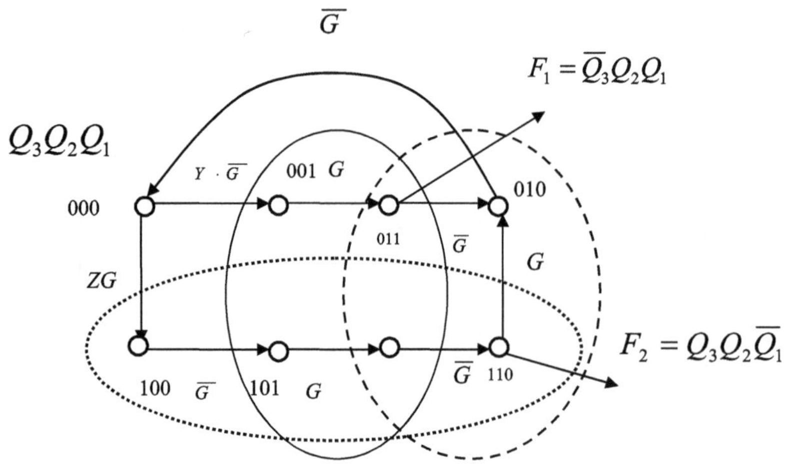

Fig. 9.17 Graph-scheme of Moore automaton

Multiple right connection, also called divergence can be used to broadcast information from its left end to each of its right ends. All ends of the connection must be of the same type.

Example 4. A finite-state machine (FSM) or finite-state automaton (plural: automata), or simply a state machine, is a mathematical abstraction sometimes used to design digital logic or computer programs. It is a behavior model composed of a finite number of states, transitions between those states, and actions, similar to a flow graph in which one can inspect the way logic runs when certain conditions are met. It has finite internal memory, an input feature that reads symbols in a sequence, one at a time without going backward; and an output feature, which may be in the form of a user interface, once the model is implemented. The operation of an FSM begins from one of the states (called a start state), goes through transitions depending on input to different states and can end in any of those available, however only a certain set of states mark a successful flow of operation (called accept states).

Finite-state machines can solve a large number of problems, among which are electronic design automation, communication protocol design, parsing and other engineering applications.

A state diagram is a type of diagram used in computer science to describe the behavior of systems. State diagrams require that the system described is composed of a finite number of states (a_i).

Build a circuit that simulates activity of a Moore automaton given by the following graph-scheme, Fig. 9.17.

Moore automata can describe the transition and output functions

$$a_{t+1} = f(a_1, x_1), z_1 = \varphi(a_1),$$

where a_1 and z_1 are the state Moore automaton and the output signal from the machine at time *t*, respectively.

The choice of memory elements. Choosing an R-S flip-flop.

Excitation function of the R-S flip-flop:

$$S_1 = \bar{Q}_3\bar{Q}_2 Y\bar{G} \vee Q_3\bar{Q}_2\bar{G},$$

$$R_1 = \bar{Q}_3 Q_2\bar{G} \vee Q_3 Q_2\bar{G} = Q_2\bar{G};$$

$$S_2 = Q_1 G,$$

$$R_2 = \bar{Q}_3\bar{Q}_1\overline{YZ};$$

$$S_3 = \bar{Q}_2\bar{Q}_1 ZG,$$

$$R_3 = Q_2\bar{Q}_1 G;$$

Output function:

$$F_1 = \bar{Q}_3 Q_2 Q_1,$$

$$F_2 = Q_3 Q_2\bar{Q}_1,$$

$$F_3 = \bar{Q}_3 Q_2 Q_1 \vee Q_3 Q_2\bar{Q}_1.$$

The process of designing large digital systems is typically one of interconnecting smaller devices, such as flip-flops, registers, multiplexers, etc., in such a way that the resulting system has the required performance characteristics.

Normally, these devices are interconnected directly to each other with various inverters being inserted as needed.

Verify the operation of the automaton we have done in the two most raspros-tranynnyh systems – NI Multisim and MicroCap 8. Diagram of the digital machine is made on triggers JK and NOR logic gates. Like NAND gates, NOR gates are so-called "universal gates" that can be combined to form any other kind of logic gate.

NI Multisim is an electronic schematic capture and simulation program which is part of a suite of circuit design programs, along with NI Ultiboard.

Micro-Cap 8 is an integrated schematic editor and mixed analog/digital simulator that provides an interactive sketch and simulates an environment for electronics engineers. Micro-Cap 8 blends a modern, intuitive interface with robust numerical algorithms.

Figure 9.18 shows a diagram of a digital automata (Micro-Cap 8).

Figure 9.19 shows the timing diagrams of a digital automata (Micro-Cap 8).

Figure 9.20 shows a diagram of a digital automata (NI Multisim 10).

Example 5. Will design an automat that will work as a binary counter when the control signal M=0 and the counter in the Gray code if M=1 (Table 9.8). The digital automata can be specified as a graph, Fig. 9.21 or Table 9.9.

The digital automata can perform as a counter, using triggers – the first version, or using multiplexers – the second version.

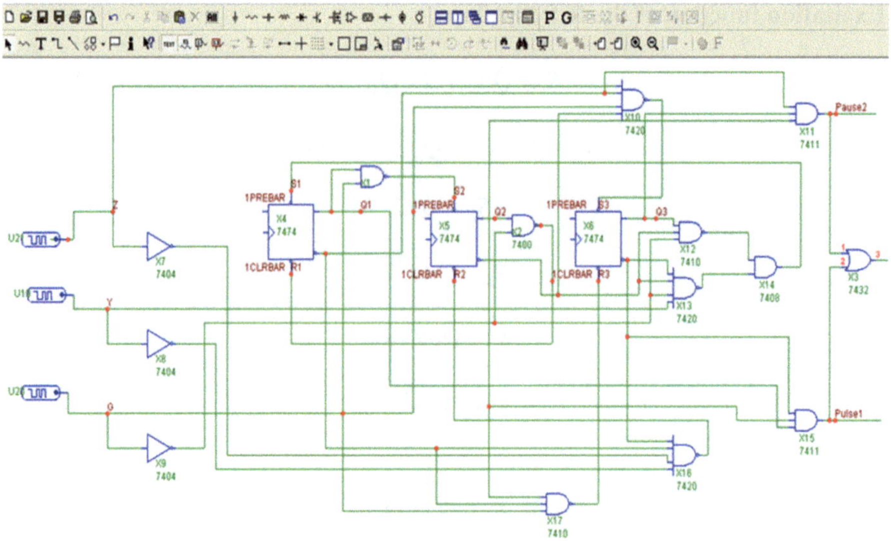

Fig. 9.18 Functional diagram of a digital automata (Micro-Cap 8)

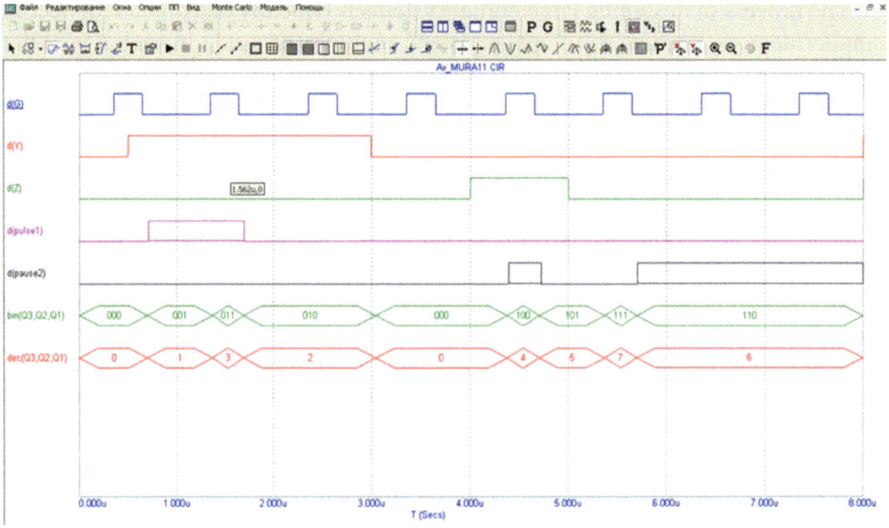

Fig. 9.19 The timing diagrams of digital automata (Micro-Cap 8)

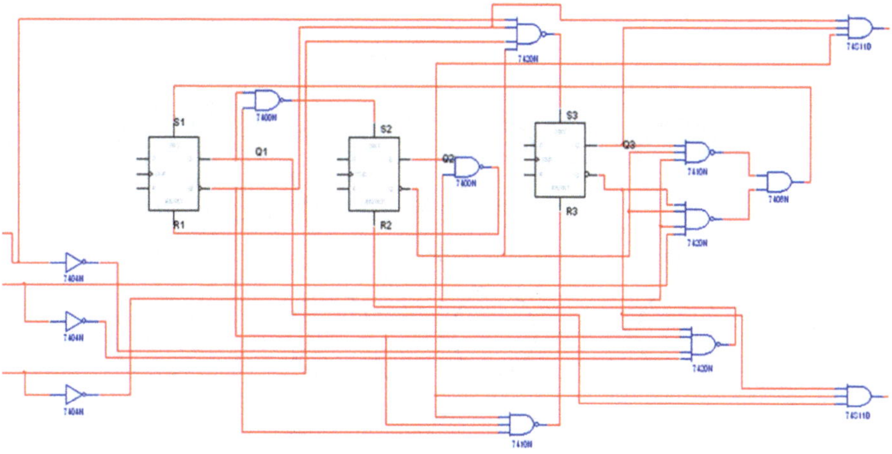

Fig. 9.20 Functional diagram of digital automata (NI Multisim 10)

Table 9.8 Gray code

Decimal	Gray code		
0	0	0	0
1	0	0	1
2	0	1	1
3	0	1	0
4	1	1	0
5	1	1	1
6	1	0	1
7	1	0	0

9.2 The First Version

Transition function for each trigger of the automaton can be described by the following dependence:

$$Q_{\mathcal{H}i} = F\left(x_1, x_2, \ldots, x_k, Q_1, Q_2, \ldots, Q_n\right).$$

Or $Q_{\mathcal{H}i} = f_i\overline{Q}_i \vee g_iQ_i$,
where Functions f and g do not contain the variables Q_i and $\overline{Q}_i$.
For J-K $Q_{\mathcal{H}i} = J_i\overline{Q}_i \vee \overline{K}_iQ_i$.
Also $fi = J_i$, $\overline{g}_i = K_i$.

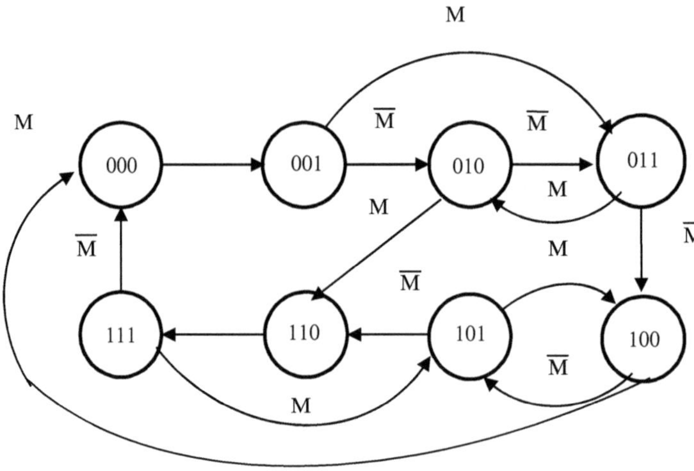

Fig. 9.21 Graph-scheme of Moore automaton

Table 9.9 The coding of states of the digital automat

Control signal	Present			Next			Control signal	Present			Next		
M	Q_2	Q_1	Q_0	Q_{2H}	Q_{1H}	Q_{0H}	**M**	Q_2	Q_1	Q_0	Q_{2H}	Q_{1H}	Q_{0H}
0	0	0	0	0	0	1	**1**	0	0	0	0	0	1
0	0	0	1	0	1	0	**1**	0	0	1	0	1	1
0	0	1	0	0	1	1	**1**	0	1	0	1	1	0
0	0	1	1	1	0	0	**1**	0	1	1	0	1	0
0	1	0	0	1	0	1	**1**	1	0	0	0	0	0
0	1	0	1	1	1	0	**1**	1	0	1	1	0	0
0	1	1	0	1	1	1	**1**	1	1	0	1	1	1
0	1	1	1	0	0	0	**1**	1	1	1	1	0	1

For $Q_i = 0$ we obtain the excitation function for the input **J**:

$$Q_{\mathcal{H}i}\Big|_{Q_i = 0} = f_i = J_i .$$

For $Q_i = 1$ we obtain the excitation function for the input **K**:

$$Q_{\mathcal{H}i}\Big|_{Q_i = 0} = g_i = \overline{K}_i .$$

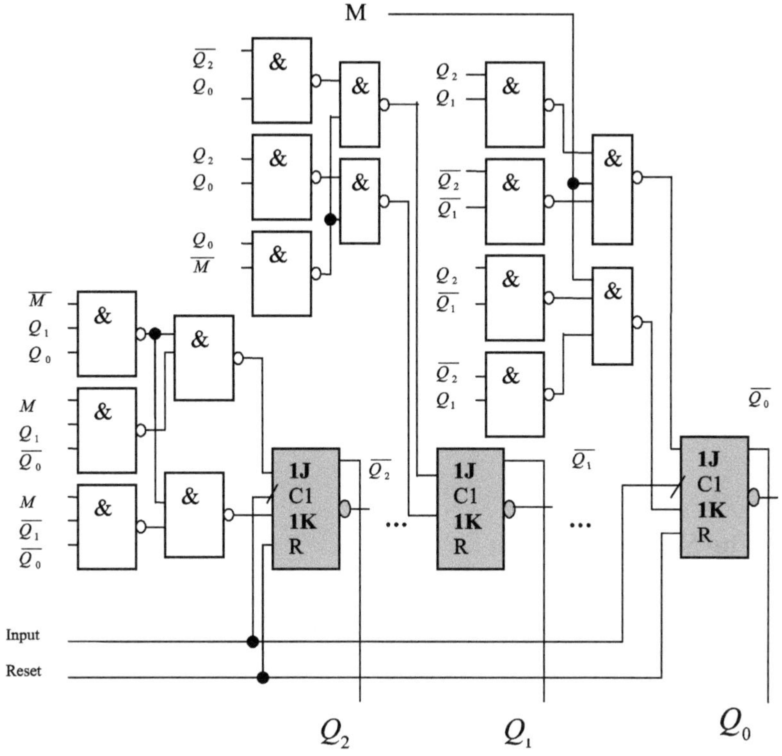

Fig. 9.22 Diagram of an automaton

Thus, we can obtain the following expressions:

$$J_2 = \bar{M}Q_1Q_0 \vee MQ_1\bar{Q}_0 = \overline{\overline{\bar{M}Q_1Q_0} \bullet \overline{MQ_1\bar{Q}_0}};$$

$$K_2 = \bar{M}Q_1Q_0 \vee M\bar{Q}_1\bar{Q}_0 = \overline{\overline{\bar{M}Q_1Q_0} \bullet \overline{M\bar{Q}_1\bar{Q}_0}};$$

$$J_1 = \bar{M}Q_0 \vee \bar{Q}_2Q_0 = \overline{\overline{\bar{M}Q_0} \bullet \overline{\bar{Q}_2Q_0}};$$

$$K_1 = \bar{M}Q_0 \vee Q_2Q_0 = \overline{\overline{\bar{M}Q_0} \bullet \overline{Q_2Q_0}};$$

$$J_0 = \bar{M} \vee \bar{Q}_2\bar{Q}_1 \vee Q_2Q_1 = \overline{M \bullet \overline{\bar{Q}_2\bar{Q}_1} \bullet \overline{Q_2Q_1}};$$

$$K_0 = \bar{M} \vee \bar{Q}_2Q_1 \vee Q_2\bar{Q}_1 = \overline{M \bullet \overline{\bar{Q}_2Q_1} \bullet \overline{Q_2\bar{Q}_1}}.$$

A diagram of an automaton is shown on Fig. 9.22.

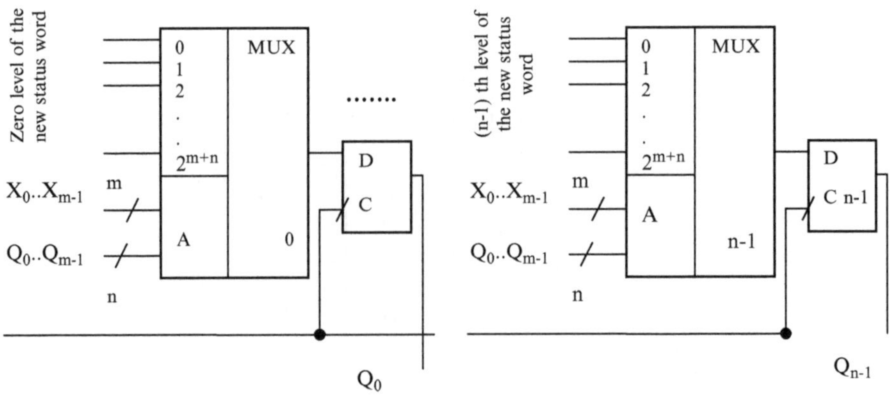

Fig. 9.23 Structure of an automaton with a multiplexer

9.3 The Second Version

9.3.1 Machine, Implemented in Flip-Flops with Multiplex Controls

Structure with multiplexers at the inputs triggers different conceptual simplicity and clarity, for its design does not require the development of logical transducers to provide the necessary transition automaton. The problem is solved, in fact, using tables. State variables are taken from the flip-flops, and input signals form a word, used as multiplexer address inputs. At this address, each multiplexer selects the variable (0 or 1) needed to transfer a D-type flip-flop to a new state.

Structure with multiplexer control triggers is shown in Figure 9.23. Inputs $x_0 \ldots x_{m-1}$ and the values of bits words of the old state $Q_0 \ldots Q_{n-1}$ form a control (address) input word multiplexer, in which the values of selected bits of the new status word are established.

For zero initial states of triggers and $M = 0$ to address inputs of multiplexers 0000 and received on the inputs of flip-flops formed by the combination of signals 001. Receipt of a clock pulse enters this combination in flip-flops. Now the address for the multiplexer is a combination of 0001, according to which they removed with a combination of 010, coming to solve the next clock pulse triggers (status register). Since the regime of binary counter. Changing the control signal M gives a regime change of the automaton. If, for example, when the word state 010 signal M becomes the unit, the address multiplexer changes from 0010 to 1010 and with their outputs show the combination of 110, corresponding to the next state when the meter is in the Gray code.

A diagram of an automaton is shown on Fig. 9.24. Dignity of the structure – easy adjustment to a new algorithm for automatic operation, the defect of – the rapid

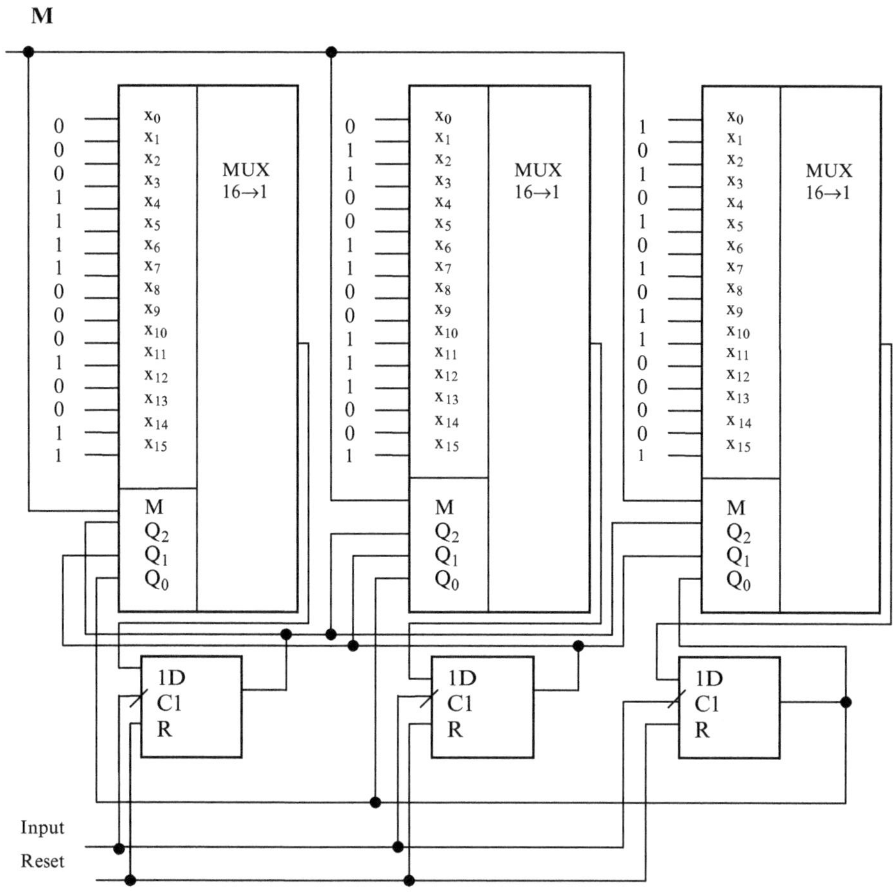

Fig. 9.24 Diagram of an automaton

growth of dimension of multiplexers with increasing number of states and inputs of the automaton.

In conclusion we would like to note that recent progress has resulted in a steady transition to digital devices with programmable logic. A programmable logic device or PLD is an electronic component used to build reconfigurable digital circuits. Unlike a logic gate, which has a fixed function, a PLD has an undefined function at the time of manufacture. Before the PLD can be used in a circuit it must be programmed, that is, reconfigured. The study of such devices is not included in the scope of this book, so we restricted ourselves to the theoretical foundations of digital automata built on their logic elements.

Appendices

Appendix A: Counter Modulo 5

Consider the implementation of counter pulses. For example solve the following problem.

Example. Develop a procedure for drawing up a decimal number to program counter modulo 5, which contains four digits. As a counter use vector $A = | a_1, a_2, a_3, a_4 |$. Schematic co unter modulo 5 (mod 5) is presented in Fig. A.1.

Counter, see Fig. A.1, assigned to count the number of pulses arriving to its input. When the next pulse counter increases the integer stored in it in 5-year notation on the 1 min value of that number – 0000, max value of that number – 4444. High bit count – 4, LBS (Least significant bit) – 1. The initial state counter 0000. In this state the counter is driven by its current value reset to 0 by feeding an impulse to tire "reset to 0" (Fig. A.2).

```
program Counter;
  {$APPTYPE CONSOLE}
  uses
   SysUtils;
// count number of digits
CONST K=4;
// variables in the program
    VAR COUNT: ARRAY [1..K] of BYTE;
    I,N:INTEGER;
// procedure reset all bits to 0
    PROCEDURE INIT;
    BEGIN
    FOR I:=1 TO K DO COUNT[I]:=0
    END;
// Function reset 1 to J of counter
 FUNCTION CHADD(J:INTEGER):BOOLEAN;
```

B. Borowik et al., *Theory of Digital Automata*, Intelligent Systems, Control and Automation: Science and Engineering 63, DOI 10.1007/978-94-007-5228-3,

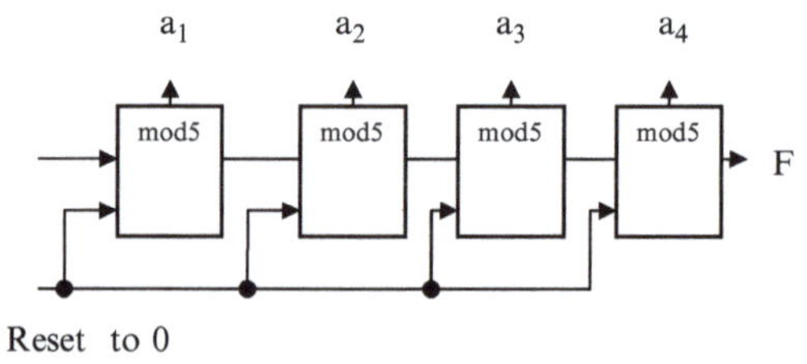

Fig. A.1 Counter modulo 5

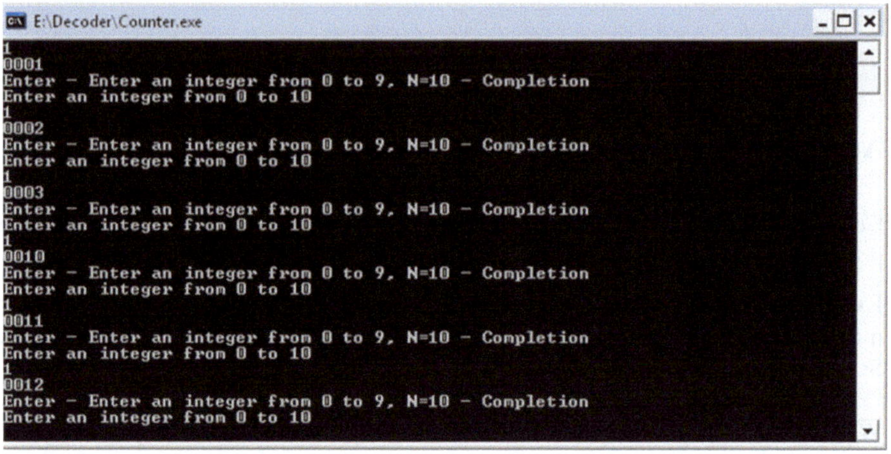

Fig. A.2 Example program "Counter"

```
   BEGIN
   CHADD:= FALSE;
   COUNT[J]:=COUNT[J]+1;
   IF COUNT[J]=K
   THEN BEGIN
   CHADD:= TRUE;
   COUNT[J]:=0 {Reset}
   END
END;
BEGIN
INIT;
REPEAT
WRITELN('Enter an integer from 0 to 10'); READLN(N);
FOR I:=1 TO N
DO IF CHADD(1)
  THEN IF CHADD(2)
  THEN IF CHADD(3)
  THEN IF CHADD(4)
THEN WRITELN('Repletion of counter. Up cast of value counter in 0');
```

```
FOR I:=K DOWNTO 1 DO WRITE(COUNT[I]:1, '');
WRITELN;
WRITELN('Enter - Enter an integer from 0 to 9, N=10 - Completion');
UNTIL N>9
  { TODO -oUser -cConsole Main : Insert code here}
end.
```

Registers

Cyclic Registers

Sometimes it is necessary to "recycle" the same values again and again. Thus the bit that usually would get dropped is fed to the register input again to receive a cyclic serial register

Figure A.3 shows the scheme register A, which contains an m + 1 bit. Originally performed by recording parallel code input bits tires in register A. The next step is a cyclical shift code combination.

Example. Develop procedures for cyclic shift right at *N* cycles code combination $A = (K_6, K_5, K_4, K_3, K_2, K_1, K_0)$. Example, from the code combination $A = (1\,0\,1\,1\,1\,0\,1)$ and $N = 8$ we should get the following results, Table A.1 and Fig. A.4.

```
program Cycle_register;
{$APPTYPE CONSOLE}
Uses SysUtils;
//MR - The number of bits of code combination
CONST MR=6;
VAR A: ARRAY[0..MR] of BYTE;
J,I,N,PZ: INTEGER;
begin
REPEAT
WRITELN('Enter (through Enter) positions of binary code combination with 6-th for 0- th');
FOR I:=MR DOWNTO 0 DO READ(A[I]);
//Enter the number of shifts
WRITELN('Conduct the number of changes to 8');
READLN(N);
WRITELN('Register A contains a code');
FOR I:=1 TO N
DO BEGIN
//Store right end digit
PZ:=A[0];
//Shifts to the right all the other bits
FOR J:=0 TO MR-1
DO A[J]:=A[J+1];
```

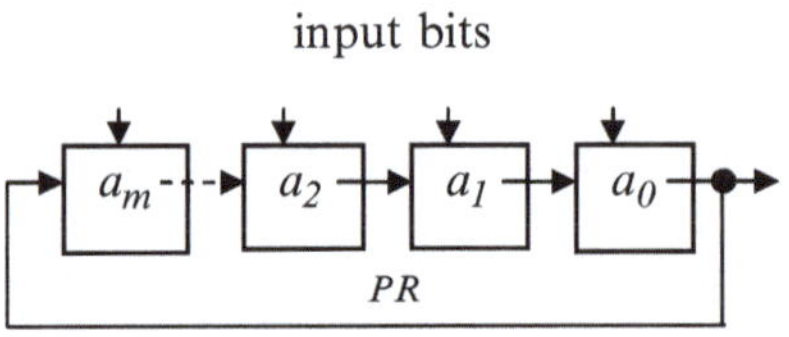

Fig. A.3 The scheme of register (A)

Table A.1 Bytes of register (A)

Step	Розряди регістру *A*
The initial state	1 0 1 1 1 0 1
Time 1	1 1 0 1 1 1 0
Time 2	0 1 1 0 1 1 1
Time 3	1 0 1 1 0 1 1
Time 4	1 1 0 1 1 0 1
Time 5	1 1 1 0 1 1 0
Time 6	0 1 1 1 0 1 1
Time 7	1 0 1 1 1 0 1
Time 8	1 1 0 1 1 1 0

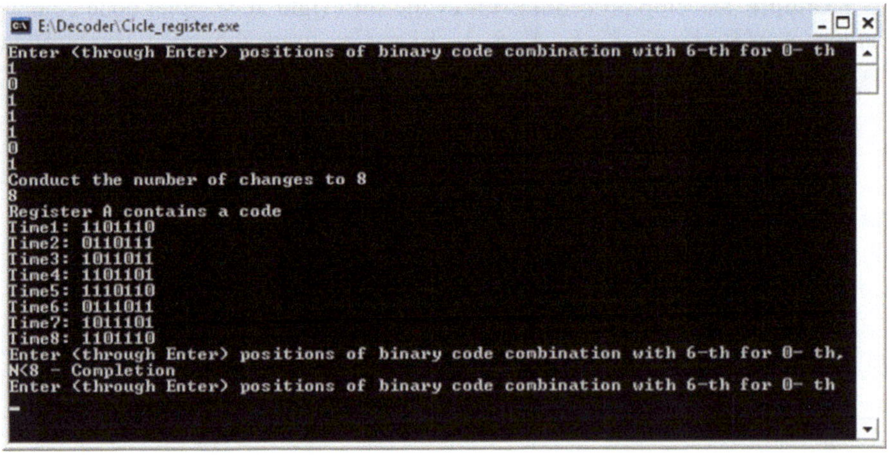

Fig. A.4 Example program "Cycle-register"

A[MR]:=PZ;
//Output code combination on each stroke of the program
WRITE('Time', I:1,': ');
FOR J:=6 DOWNTO 0 DO WRITE(A[J]:1, '');
WRITELN;
END;
WRITELN('Enter (through Enter) positions of binary code combination with 6-th for 0- th, N<8 - Completion');

Fig. A.5 Scheme of binary coder

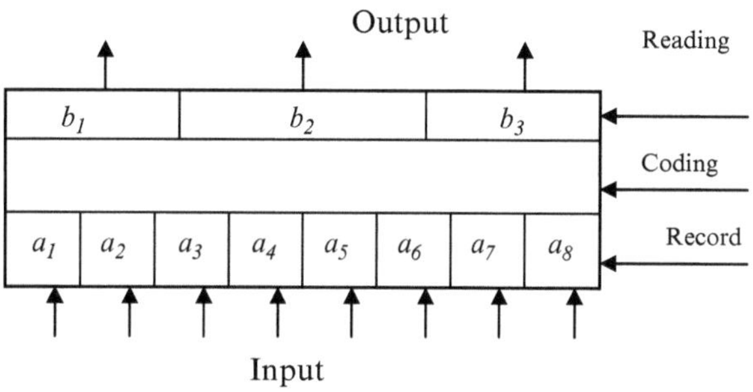


```
UNTIL N<8
{TODO -oUser -cConsole Main : Insert code here}
end.
```

Binary Coder

Example. Develop a procedure for encoding binary position 8-bit binary code A in the three-digit code B.

To store the position of 8-bit binary code to use vector – A:

$$A = |a_1, a_2, a_3, a_4, a_5, a_6, a_7, a_8|.$$

To store binary 3-bit code to use vector – B:

$$B = |b_1, b_2, b_3|.$$

Schematic encoder shown in Fig. A.5.

For an illustration of the coder we discuss some examples of coding:

- Let A = (10000000), then for this code B = (000);
- Let A = (01000000), then for this code B = (001);
- Let A = (00100000), then for this code B = (010);
- Let A = (00010000), then for this code B = (011);
- Let A = (00001000), then for this code B = (100);
- Let A = (00000100), then for this code B = (101);
- Let A = (00000010), then for this code B = (110);
- Let A = (00000001), then for this code B = (111).

Based on these options we can make a code table (see Table A.2).

Based on Table A.2 we can create a logical encoding scheme (Figs. A.6 and A.7). The scheme employs three logical gates "**OR**" (the 3-bit in the code B).

```
program Coder;
{$APPTYPE CONSOLE}
uses
```

Table A.2 Code table of binary coder

	Code A	Code B		
		b_1	b_2	b_3
a_1	10000000			
a_2	01000000			+
a_3	00100000		+	
a_4	00010000		+	+
a_5	00001000	+		
a_6	00000100	+		+
a_7	00000010	+	+	
a_8	00000001	+	+	+

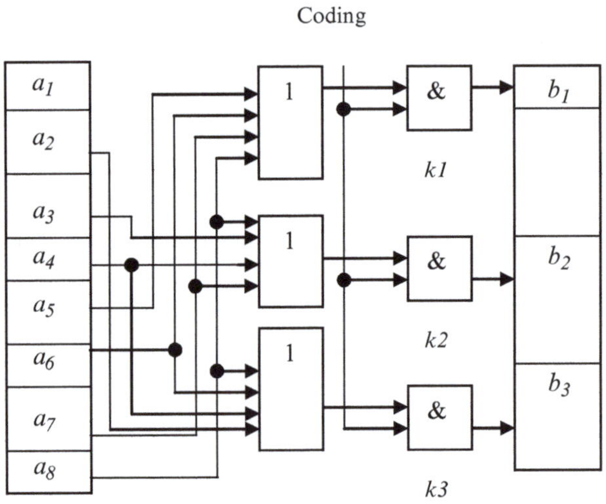

Fig. A.6 The scheme of binary coder employ three logic gates "OR"

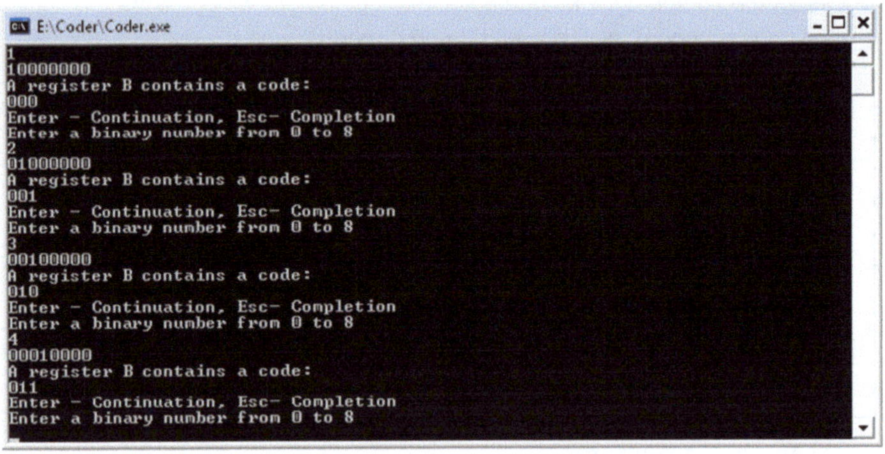

Fig. A.7 Example program "Binary coder" (Pascal)

```
SysUtils;
VAR A: ARRAY[1..8] of BYTE;
     B: ARRAY[1..3] of BYTE;
     K1, K2, K3:BOOLEAN;
     i,n:INTEGER;
BEGIN
  REPEAT
  WRITELN('Enter a binary number from 0 to 8');
  READLN(N);
  FOR I:=1 TO 8
//form a binary code number of positional
  DO IF I = N THEN A[I]:= 1
  ELSE A[I]:= 0;
  FOR I:=1 TO 8 DO WRITE(A[I]:1, '');
  WRITELN;
// Output Register A
  K1:= (A[5]=1) OR (A[6]=1) OR (A[7]=1) OR (A[8]=1);
  K2:= (A[3]=1) OR (A[6]=1) OR (A[7]=1) OR (A[8]=1);
  K3:= (A[2]=1) OR (A[4]=1) OR (A[6]=1) OR (A[8]=1);
// Record in case in Register B
  IF K1 THEN B[1]:=1 ELSE B[1]:=0;
  IF K2 THEN B[2]:=1 ELSE B[2]:=0;
  IF K3 THEN B[3]:=1 ELSE B[3]:=0;
  WRITELN ('A register B contains a code:');
  FOR I:=1 TO 3 DO WRITE(B[1]:1, ''); WRITELN;
 WRITELN('Enter - Enter a binary number from 0 to 8, 9 - Completion');
  UNTIL n>8
 {TODO -oUser -cConsole Main : Insert code here}
end.
```

The following is a listing in C#, describing the work of the encoder (Fig. A.8).

```
#include <iostream>
using namespace std;
int main()
{
    int a[8];
    int b[3];
    bool k1,k2,k3;
    int i,n;
    cout<<"Enter a binary number from 0 to 8\n";
    cin>>n;
    while(n!=9)
    {
    for(int i=0;i<8;i++)
```

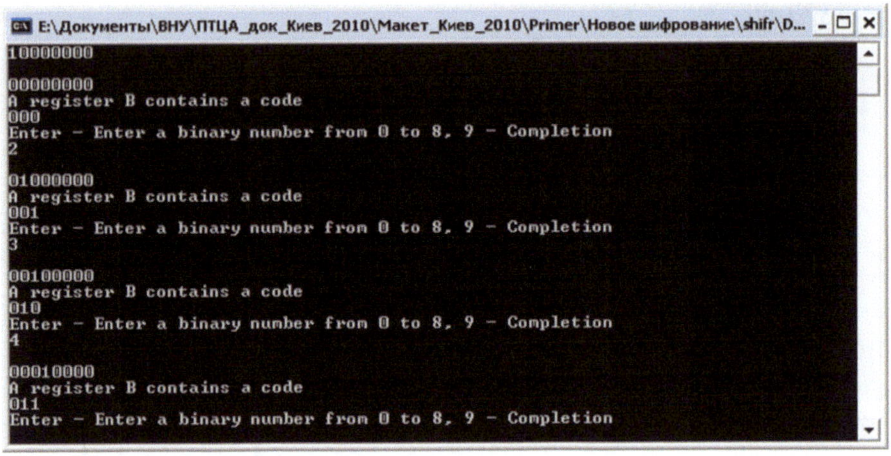

Fig. A.8 Example program "Binary coder" (C#)

```
{
   if(i+1==n)
   {
      a[i]=1;
   }
   else
   {
      a[i]=0;
   }
}
cout<<"\n";
for(int i=0;i<8;i++)
{
cout<<a[i];
}
cout<<"\n";
k1=(a[4]==1)|(a[5]==1)|(a[6]==1)|(a[7]==1);
k2=(a[2]==1)|(a[3]==1)|(a[6]==1)|(a[7]==1);
k3=(a[1]==1)|(a[3]==1)|(a[5]==1)|(a[7]==1);
if(k1)
{
   b[0]=1;
}
else
{
   b[0]=0;
}
if(k2)
```

```
{
   b[1]=1;
}
else
{
   b[1]=0;
}
if(k3)
{
   b[2]=1;
}
else
{
   b[2]=0;
}
cout<<"A register B contains a code\n";
for(int i=0;i<3;i++)
{
   cout<<b[i];
}
cout<<"\n";
cout<<"Enter - Enter a binary number from 0 to 8, 9 - Completion\n";
cin>>n;
}
}
```

Example encoder on a chip 74147 (MC8) is shown in Fig. A.9.

Binary Decoder

Example. Develop a procedure for decoding the 3 bit binary code *A* in binary positional 8-bit code *B*.

To store binary 3-bit code to use vector – *A*:

$$A = |a_1, a_2, a_3|.$$

To store the position of 8-bit binary code to use vector – *B*:

$$B = |b_1, b_2, b_3, b_4, b_5, b_6, b_7, b_8|.$$

A schematic decoder is shown on Fig. A.10.

For an illustration of the decoder we discuss some examples of decoding:

- Let *A* = (000), then for this code *B* = (10000000);
- Let *A* = (001), then for this code *B* = (01000000);
- Let *A* = (010), then for this code *B* = (00100000);

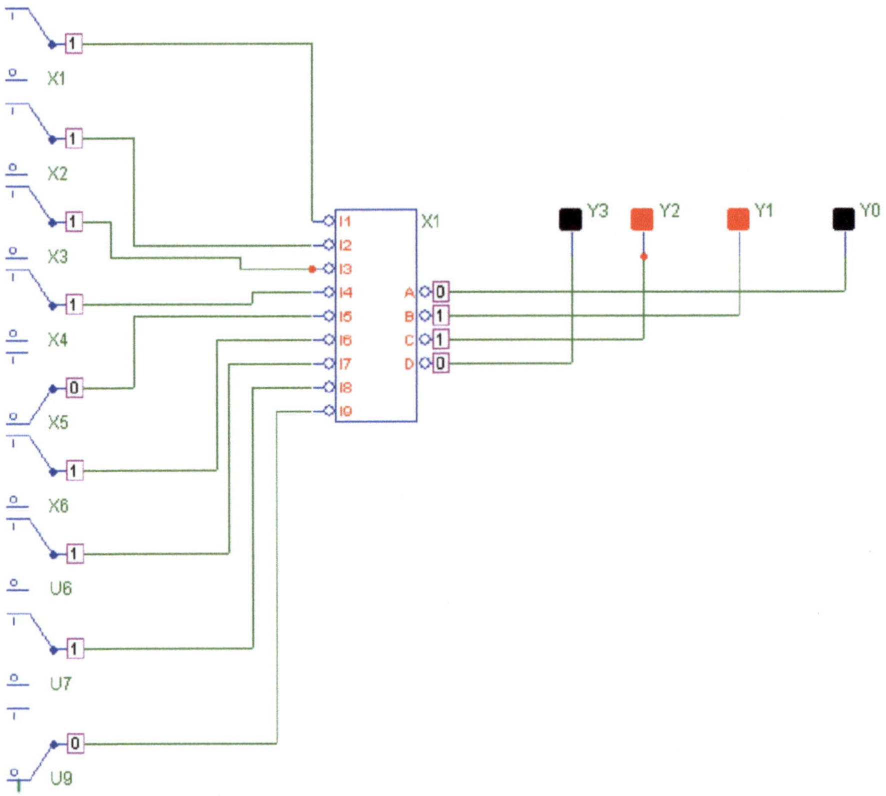

Fig. A.9 Example of encoder on the 74147 chip (MC8)

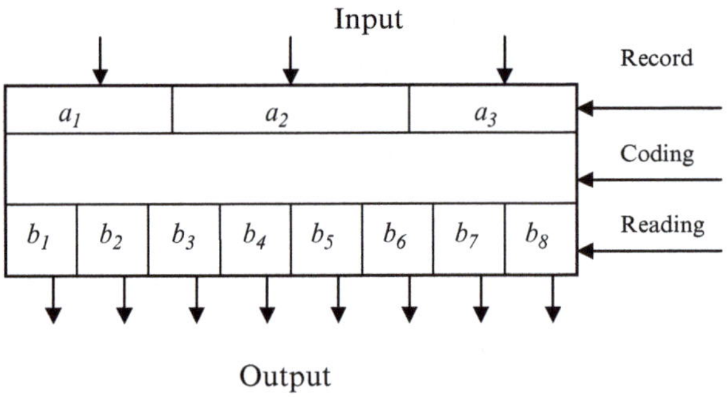

Fig. A.10 Scheme of binary decoder

Table A.3 Code table of binary decoder

Code A	Code *B*							
(a_1, a_2, a_3)	b_1	b_2	b_3	b_4	b_5	b_6	b_7	b_8
0 0 0	+							
0 0 1		+						
0 1 0			+					
0 1 1				+				
1 0 0					+			
1 0 1						+		
1 1 0							+	
1 1 1								+

- Let $A = (011)$, then for this code $B = (00010000)$, etc.

Based on these options we can make a code table (see Table A.3).
Enter the following logical variables:

$$1 = \overline{a1} \wedge \overline{a2} \wedge \overline{a3}; \quad d2 = \overline{a1} \wedge \overline{a2} \wedge a3; \; d3 = \overline{a1} \wedge a2 \wedge \overline{a3};$$

$$d4 = \overline{a1} \wedge a2 \wedge a3; \quad d5 = a1 \wedge \overline{a2} \wedge \overline{a3}; \; d6 = a1 \wedge \overline{a2} \wedge a3;$$

$$d7 = a1 \wedge a2 \wedge \overline{a3}; \quad d8 = a1 \wedge a2 \wedge a3.$$

Based on these options we can make a code table (see Table A.3).
Logical circuit decoding is shown in Figs. A.11 and A.12.

```
program Decoder;
{$APPTYPE CONSOLE}
Uses SysUtils;
      VAR B: ARRAY[1..8] of BYTE; A: ARRAY[1..3] of BYTE;
      D: ARRAY[1..8] of BOOLEAN;
      i,N:INTEGER;
  BEGIN
  REPEAT
  WRITELN('Enter through blanks in a register "A" a three-digit binary number');
  READLN(A[1],A[2],A[3]);
  // Decoding
  D[1]:=(A[1]=0) and (A[2]=0) and (A[3]=0);
  D[2]:=(A[1]=0) and (A[2]=0) and (A[3]=1);
  D[3]:=(A[1]=0) and (A[2]=1) and (A[3]=0);
  D[4]:=(A[1]=0) and (A[2]=1) and (A[3]=1);
  D[5]:=(A[1]=1) and (A[2]=0) and (A[3]=0);
  D[6]:=(A[1]=1) and (A[2]=0) and (A[3]=1);
  D[7]:=(A[1]=1) and (A[2]=1) and (A[3]=0);
  D[8]:=(A[1]=1) and (A[2]=1) and (A[3]=1);
```

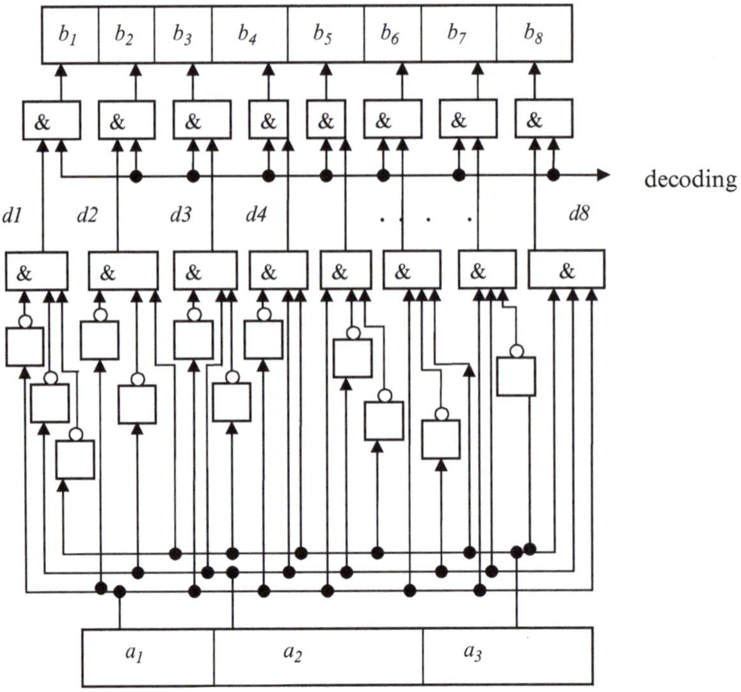

Fig. A.11 Example program "Binary coder" (C#)

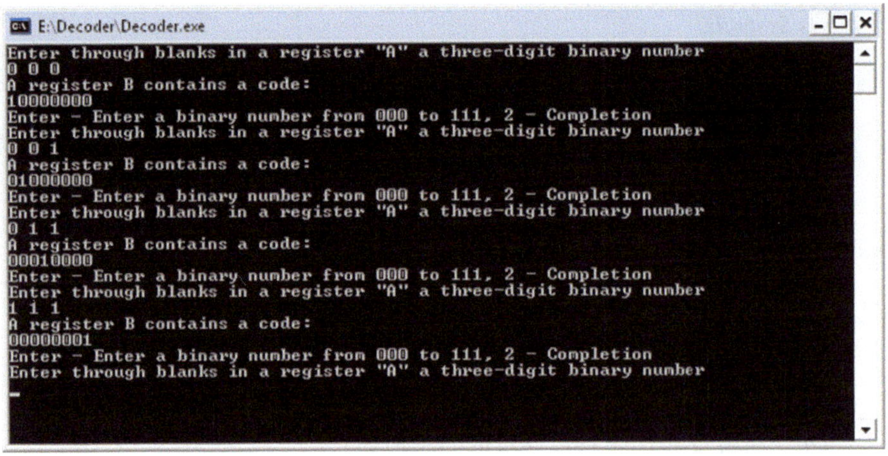

Fig. A.12 Example program "Binary decoder" (Pascal)

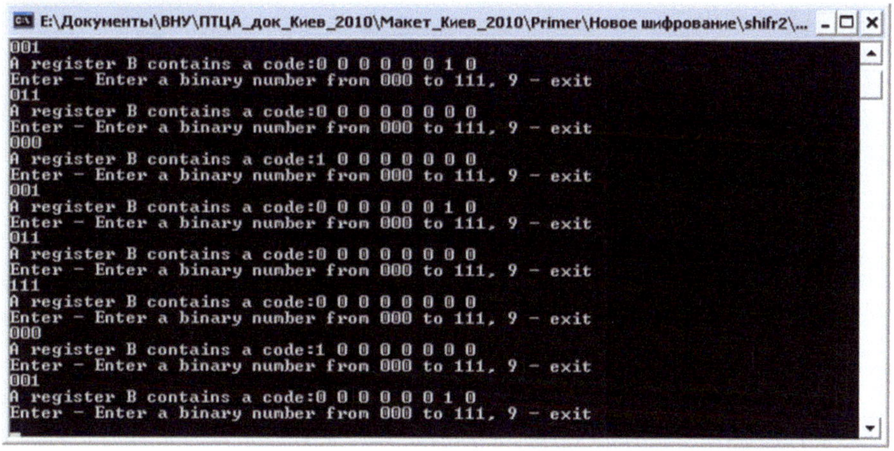

Fig. A.13 Example program "Binary decoder" (C#)

```
// Output Register B
    FOR I:=1 TO 8 DO IF D[I]
    THEN B[I]:=1 ELSE B[I]:=0;
    WRITELN ('A register B contains a code:');
    FOR I:=1 TO 8 DO WRITE(B[I]:1, ''); WRITELN;
     WRITELN('Enter - Enter a binary number from 000 to 111, n>9
Completion');
    UNTIL n>9
    {TODO -oUser -cConsole Main : Insert code here}
   end.
```

The following is a listing in C#, describing the work of the binary decoder (Fig. A.13).

```
#include<iostream>
#include "stdio.h"
using namespace std;
void main()
{
   int c,n;
   int b[8];
   int a[3];
   bool D[8];
   char str[6];
   char *px1,*px2,*px3;
   cout<<"Enter through blanks in a register 'A' a three-digit binary number\n";
   gets(str);
   while(str[0]!='9')
   {
```

```
        px1=&str[0];
        px2=&str[2];
        px3=&str[4];
        a[0]=atoi(px1);
        a[1]=atoi(px2);
        a[2]=atoi(px3);
        D[0]=(a[0]==0)&(a[1]==0)&(a[2]==0);
        D[1]=(a[0]==0)&(a[1]==0)&(a[2]==1);
        D[2]=(a[0]==0)&(a[1]==1)&(a[2]==0);
        D[3]=(a[0]==0)&(a[1]==1)&(a[2]==1);
        D[4]=(a[0]==1)&(a[1]==0)&(a[2]==0);
        D[5]=(a[0]==1)&(a[1]==0)&(a[2]==1);
        D[6]=(a[0]==1)&(a[1]==1)&(a[2]==0);
        D[7]=(a[0]==1)&(a[1]==1)&(a[2]==1);
        for(int i=0;i<8;i++)
        {
          if(D[i])
          {
            b[i]=1;
          }
          else
          {
            b[i]=0;
          }
        }
        cout<<"A register B contains a code:";
        for(int i=0;i<8;i++)
        {
          cout<<b[i]<<' ';
        }
        cout<<endl;
        cout<<"Enter - Enter a binary number from 000 to 111, 9 - exit\n";
        gets(str);
        }
      }
```

Example of the decode binary code on the seven segment LED IC 7448 (MC8) is shown in Fig. A.14.

Appendix B: Full Adder Circuit

A Full Adder is a combinational circuit that performs the arithmetic sum of three input bits. It consists of three inputs and two outputs. Three of the input variables can be defined as A, B, Cin and the two output variables can be defined as S, Cout.

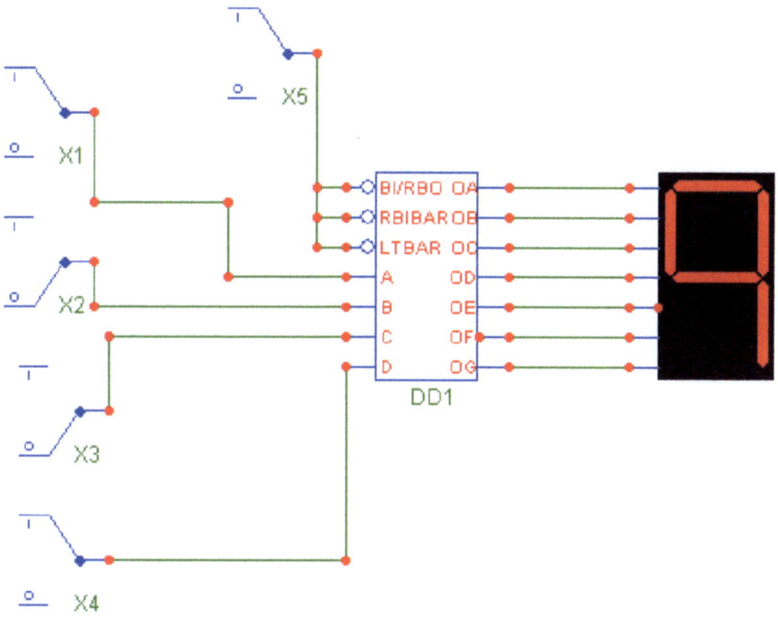

Fig. A.14 Example of the decoding binary code on the seven segment LED IC 7448 (MC8)

The two input variables that we defined earlier A and B represents the two significant bits to be added. The third input Cin represents the carry bit. We have to use two digits because the arithmetic sum of the three binary digits needs two digits. The two outputs represents S for sum and Cout for carry.

For designing a full adder circuit, two half adder circuits and an OR gate is required. It is the simplest way to design a full adder circuit. For this two XOR gates, two AND gates, one OR gate is required.

The logic circuit in a full adder MicroCap 8 (MC8).
Fig. A.15

In order to implement the serial adder, it is necessary to use some device capable of storing the information regarding the presence or absence of a carry. Such a device must have two distinct states, such that each can be assigned to represent a state of the adder. A number of such devices exist, among which is the delay element, which may simply consist of a D flip-flop.

Adder at IC 7448 (Fig. A.16).

The timing diagrams of the adder (IC 7448) (Fig. A.17).

Appendix C

Step 1. Circuit of the counter on J-K flip-flop (Fig. A.18).

Step 2. Circuit of the counter on IC 4027, IC 4069, IC 4081, IC 4511 (Fig. A.19).

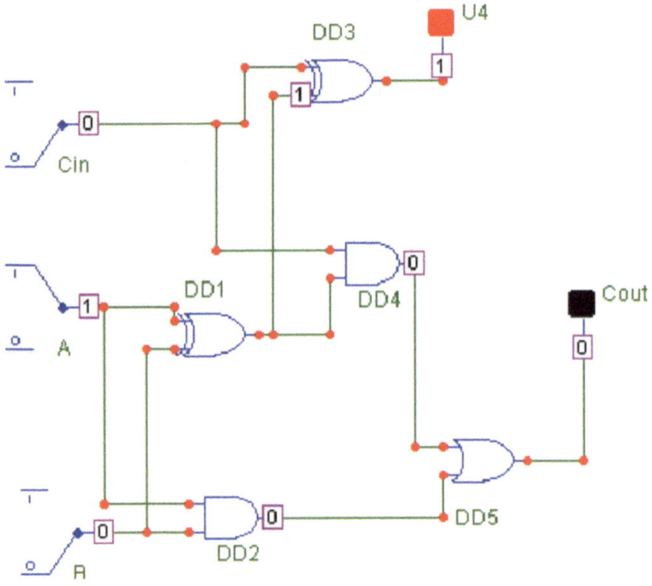

Fig. A.15 Full Adder Circuit (MicroCap 8)

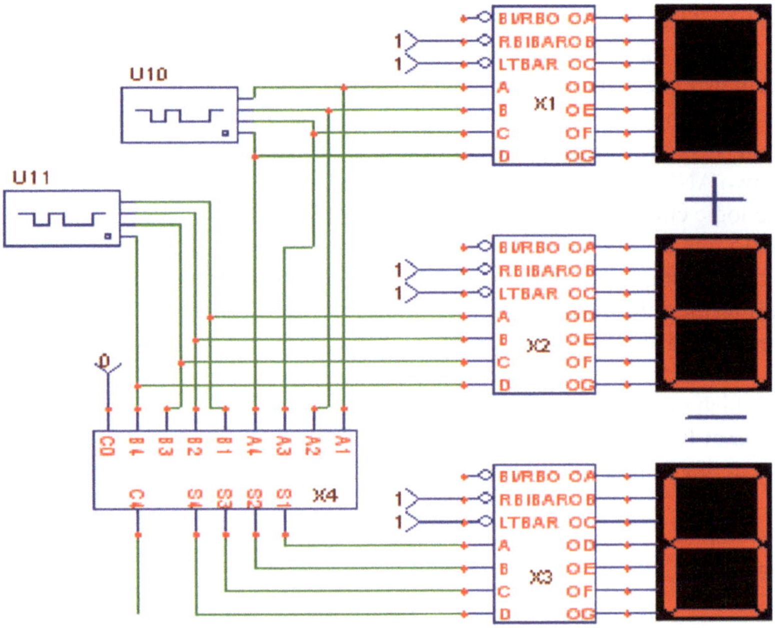

Fig. A.16 BCD Adder using IC 7448

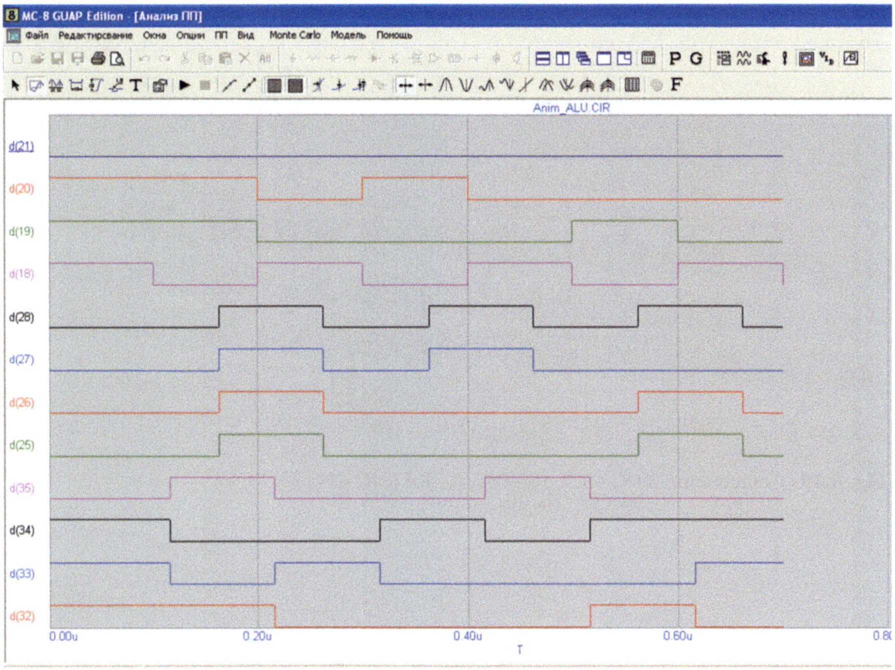

Fig. A.17 Timing diagram of the full adder (IC 7448)

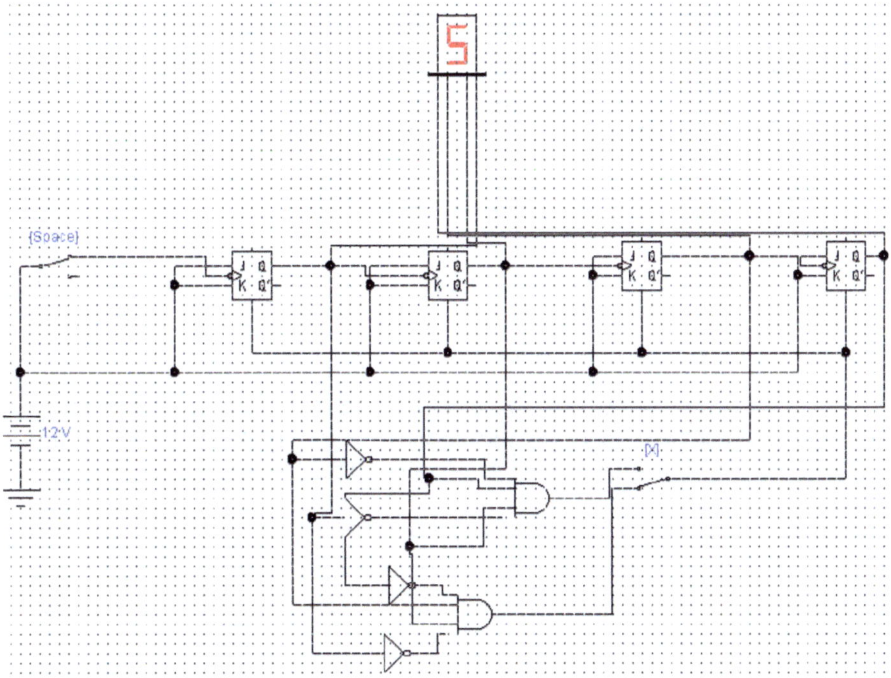

Fig. A.18 JK flip-flop counter circuits

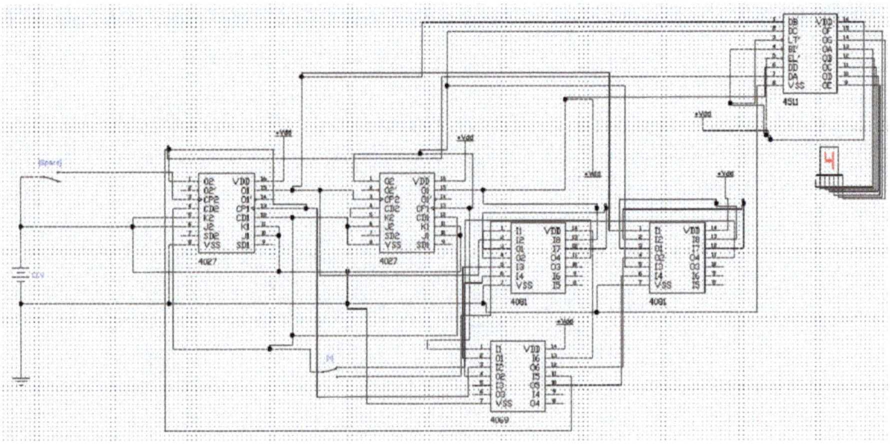

Fig. A.19 Counter circuit (IC 4027, IC 4069, IC 4081, IC 4511)

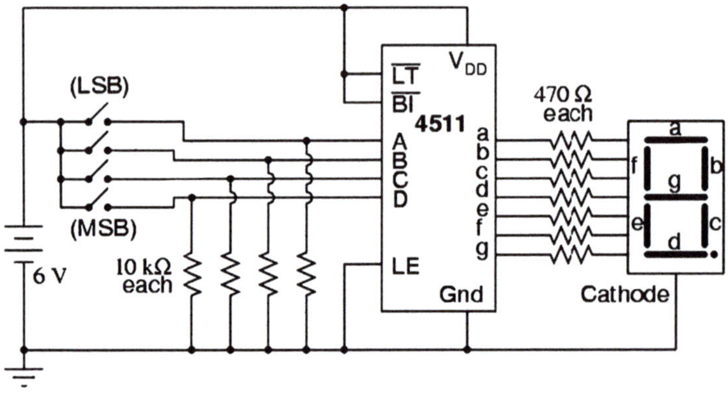

Fig. A.20 Schematic diagram of the connection IC 4511

PHOTO-RESIST BOARD is a piece of glass reinforce plastic. One of the sides is copper clad and this copper has a photosensitive coating. When the plastic film is peeled back this sensitive coating is revealed. After processing this will be the PCB.

Schematic diagram of the connection IC 4511 (Fig. A.20).

Step 3. Printed Wiring Board (PWB) was developed in the program will DipTrace (Fig. A.21).

Step 4. Manufacturing of Printed Wiring Board (Figs. A.22 and A.23).

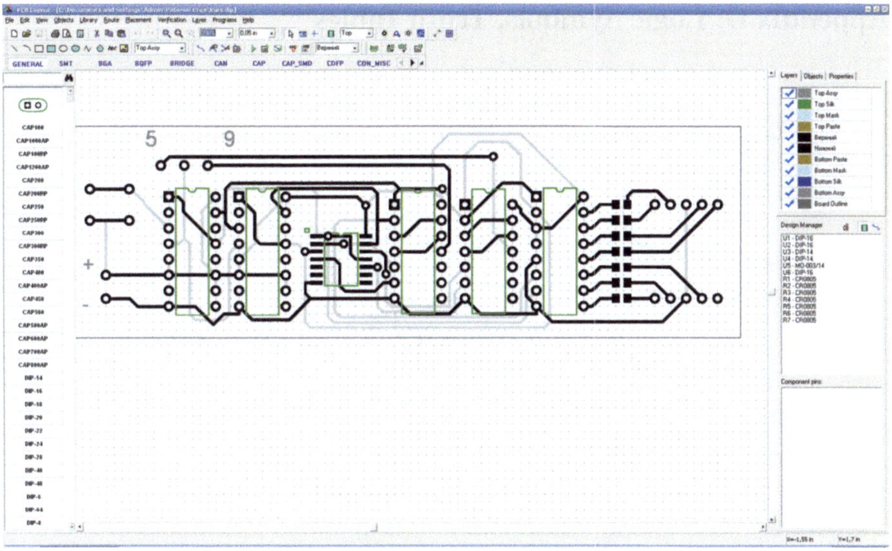

Fig. A.21 Circuit diagram to PCB

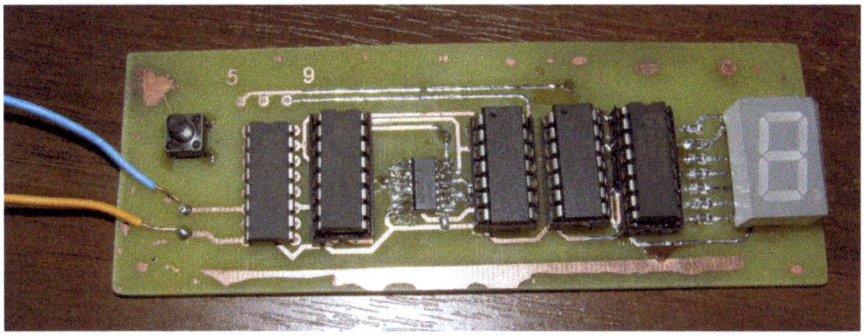

Figs. A.22 Printed circuit board (Side 1)

Fig. A.23 Printed circuit board (Side 2)

Appendix D: Logic Symbols, Truth Tables

AND	OR	A	B	Y
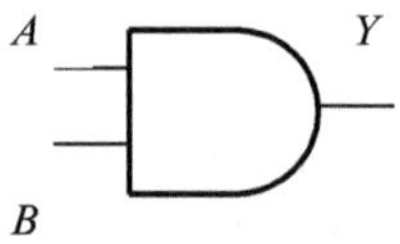	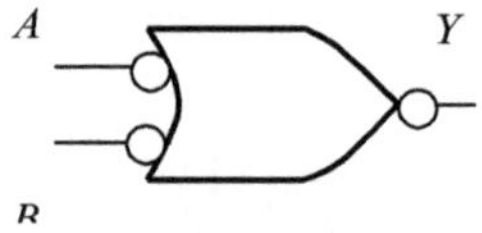	1 1 0 0	1 0 1 0	1 0 0 0
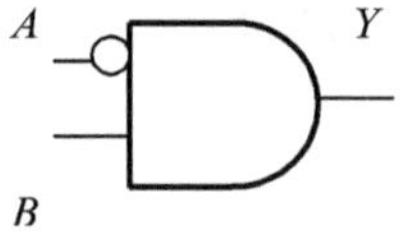	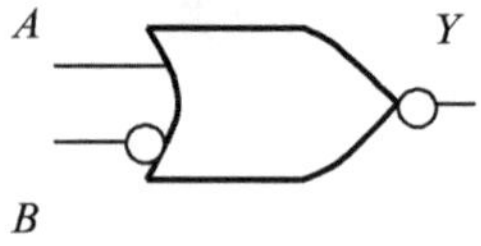	1 1 0 0	1 0 1 0	0 0 1 0
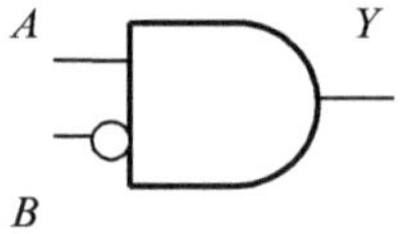	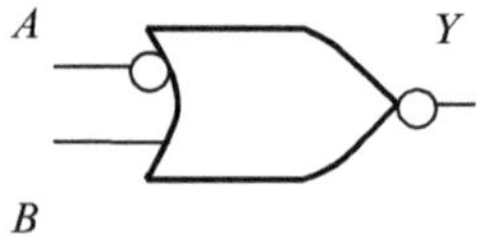	1 1 0 0	1 0 1 0	0 1 0 0
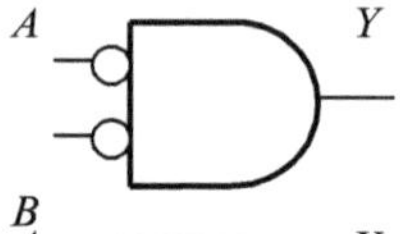	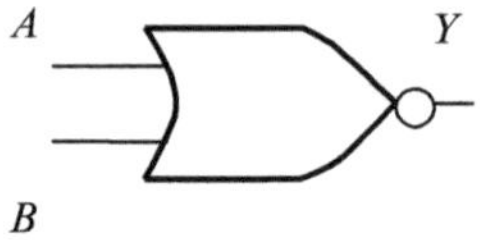	1 1 0 0	1 0 1 0	0 0 0 1
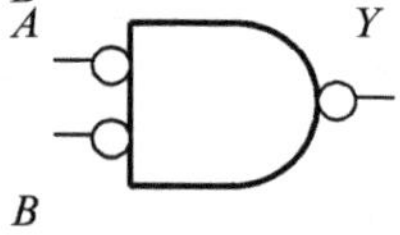	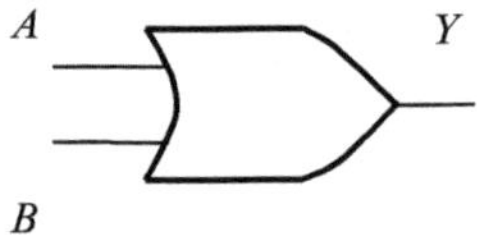	1 1 0 0	1 0 1 0	1 1 1 0
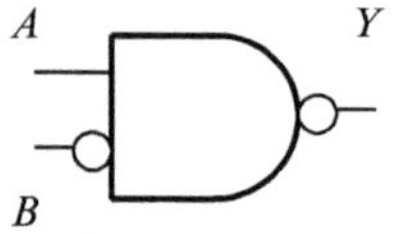	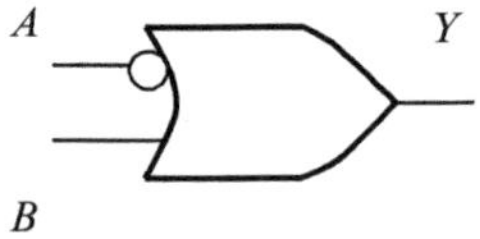	1 1 0 0	1 0 1 0	1 0 1 1
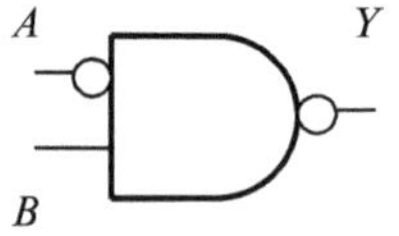	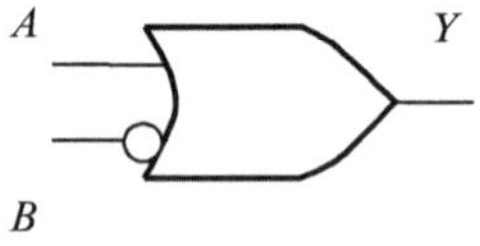	1 1 0 0	1 0 1 0	1 1 0 0
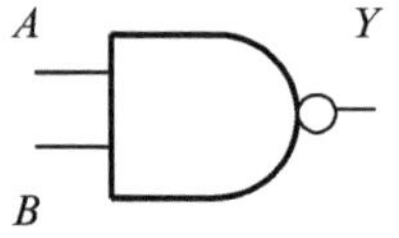	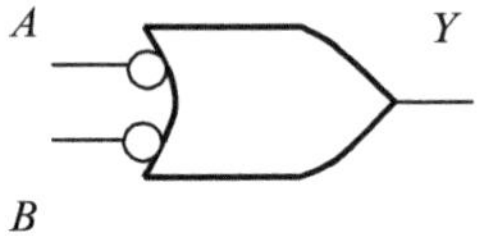	1 1 0 0	1 0 1 0	0 1 1 1

Glossary

A

Access time The time from the application of a valid memory address to the appearance of valid output data

Address The location of a given storage cell or group of cells in a memory; a unique memory location containing 1 byte

Adjacency Characteristic of cells in a Karnaugh map in which there is a single-variable change from one cell to another cell next to it on any of its four sides

Alphanumeric Consisting of numerals, letters, and other characters

ALU Arithmetic logic unit; the key processing element of a microprocessor that performs arithmetic and logic operations

Amplitude In a pulse waveform, the height or maximum value of the pulse as measured from its LOW level

Analog Being continuous or having continuous values, as opposed to having a set of discrete values

Analog-to-digital (A/D) conversion The process of converting an analog signal to digital form

Analog-to-digital converter (ADC) A device used to convert an analog signal to a sequence of digital codes

AND A basic logic operation in which a true (HIGH) output occurs only when all the input conditions are true (HIGH)

AND gate A logic gate that produces a HIGH output only when all of the inputs are HIGH

ANSI American National Standards Institute

ASCII American Standard Code for Information Interchange; the most widely used alphanumeric code

Associative law In addition (ORing) and multiplication (ANDing) of three or more variables, the order in which the variables are grouped makes no difference

Asynchronous counter A type of counter in which each stage is clocked from the output of the preceding stage

B. Borowik et al., *Theory of Digital Automata*, Intelligent Systems, Control and Automation: Science and Engineering 63, DOI 10.1007/978-94-007-5228-3,

B

Base One of the three regions in a bipolar junction transistor
Base address The beginning address of a segment of memory
BCD Binary coded decimal; a digital code in which each of the decimal digits, 0–9, is represented by a group of 4 bits
Binary Having two values or states; describes a number system that has a base of 2 and utilizes 1 and 0 as its digits
Bipolar Having two opposite charge carriers within the transistor structure
Bistable Having two stable states. Flip-flops and latches are bistable multivibrators
Bit A binary digit, which can be either a 1 or 0
Boolean addition In Boolean algebra, the OR operation
Boolean algebra The mathematics of logic circuits
Boolean expression An expression of variables and operators used to express the operation of a logic circuit
Boolean multiplication In Boolean algebra, the AND operation
Byte A group of 8 bits

C

Cache memory A relatively small. high-speed memory that stores the most recently used instructions or data from the larger but slower main memory
Capacity The total number of data units (bits, bytes, words) that a memory can store
Cascade To connect "end-to-end" as when several counters are connected from the terminal count output of one counter to the enable input of the next counter
Cascading Connecting the output of one device to the input of a similar device, allowing one device to drive another in order to expand the operational capability
CCD Charge-coupled device: a type of semiconductor memory that stores data in the form of charge packets and is serially accessed
Cell An area on a Karnaugh map that represents a unique combination of variables in product form: a single storage element in a memory
CMOS Complementary metal oxide semiconductor; a class of integrated logic circuits that is implemented with a type of field-effect transistor
Combinational logic A combination of logic gates interconnected to produce a specified Boolean function with no storage or memory capability: sometimes called combinatorial logic
Commutative law In addition (ORing) and multiplication (AND-ing) of two variables, the order in which the variables are OR or AND makes no difference
Comparator A digital circuit that compares the magnitudes of two quantities and produces an output indicating the relationship of the quantities

Counter A digital circuit capable of counting electronic events, such as pulses by progressing through a sequence of binary states

CPLD A complex programmable logic device that consists basically of multiple SPLD arrays with programmable interconnections

CPU Central processing unit: the main part of a computer responsible for control and processing of data: the core of a DSP that processes the program instructions

D

Data Information in numeric, alphabetic, or other form

Decade counter A digital counter having ten states

Decimal Describes a number system with a base of ten

Decoder A digital circuit that converts coded information into another (familiar) or noncoded form

D flip-flop A type of bistable multivibrator in which the output assumes the state of the D input on the triggering edge of a clock pulse

Demultiplexer (demux) A circuit (digital device) that switches digital data from one input line to several output lines in a specified time sequence

Dependency notation A notational system for logic symbols that specifies input and output relationships, thus fully defining a given function; an integral feature of ANSI/IEEE Std. 91-1984

Digital-to-analog (D/A) conversion The process of converting a sequence of digital codes to an analog form

Digital-to-analog converter (DAC) A device in which information in digital form is converted to analog form

DIP Dual in-line package; a type of IC package whose leads must pass through holes to the other side of a PC board

Distributive law The law that states that ORing several variables and then ANDing the result with a single variable is equivalent to ANDing the single variable with each of the several variables and then ORing the product

Domain All of the variables in a Boolean expression

"Don't care" A combination of input literals that cannot occur and can be used as a 1 or a 0 on a Karnaugh map for simplification

E

Edge-triggered flip-flop A type of flip-flop in which the data are entered and appear on the output on the same clock edge

Emitter One of the three regions in a bipolar junction transistor

Enable To activate or put into an operational mode; an input on a logic circuit that enables its operation

Exclusive-NOR (XNOR) gate A logic gate that produces a LOW only when the two inputs are at opposite levels
Exclusive-OR (XOR) A basic logic operation in which a HIGH occurs when the two inputs are at opposite levels
Exclusive-OR (XOR) gate A logic gate that produces a HIGH only when the two inputs are at opposite levels

F

Flip-flop A basic storage circuit that can store only one bit at a time; a synchronous bistable device
Floating-point number A number representation based on scientific notation in which the number consists of an exponent and a mantissa

G

Gate A circuit having two or more input terminals and one output terminal, where an output is present only when the prescribed inputs are present
Gray code A cyclic code, similar to a binary code, in which only one bit changes as the counting number increases

H

Hamming code A type of error-correction code
Handshaking The process of signal interchange by which two digital devices or systems jointly establish communication
HDL Hardware description language; a language used for describing a logic design using software
Hexadecimal Describes a number system with a base of 16

I

IEEE Institute of Electrical and Electronics Engineers
Input device Any connected equipment, such as digital control devices or peripheral devices, that supply information to the central processing unit. Each type of input device has a unique interface to the processor

J

J-K flip-flop A type of flip-flop that can operate in the SET, RESET, no-change, and toggle modes

Johnson counter A type of register in which a specific prestored pattern of 1s and 0s is shifted through the stages, creating a unique sequence of bit patterns

K

Karnaugh map An arrangement of cells representing the combinations of literals in a Boolean expression and used for a systematic simplification of the expression

L

Literal A variable or the complement of a variable

Load To enter data into a shift register

Logic In digital electronics, the decision-making capability of gate circuits, in which a HIGH represents a true statement and a LOW represents a false one

Logic element The smallest section of logic in an FPGA that typically contains an LUT. associated logic, and a flip-flop

M

Machine code The basic binary instructions understood by the processor

Mantissa The magnitude of a floating-point number

Minimization The process that results in an SOP or POS Boolean expression that contains the fewest possible terms with the fewest possible literals per term

Multiplexer (mux) A circuit (digital device) that switches digital data from several input lines onto a single output line in a specified time sequence

Multivibrator A class of digital circuits in which the output is connected back to the input (an arrangement called feedback) to produce either two stable states, one stable state, or no stable states, depending on the configuration

N

NAND gate A logic circuit in which a LOW output occurs only if all the inputs are HIGH

Negative-AND An equivalent NOR gate operation in which the HIGH is the active input when all inputs are LOW
Negative-OR An equivalent NAND gate operation in which the HIGH is the active input when one or more of the inputs are LOW
Netlist A detailed listing of information necessary to describe a circuit, such as types of elements, inputs, and outputs, and all interconnections
Nibble A group of 4 bits
NMOS An *n*-channel-metal-oxide semiconductor
Node A common connection point in a circuit in which a gate output is connected to one or more gate inputs
Noise immunity The ability of a circuit to reject unwanted signals
Noise margin The difference between the maximum LOW output of a gate and the maximum acceptable LOW input of an equivalent gate; also, the difference between the minimum HIGH output of a gate and the minimum HIGH input of an equivalent gate
Nonvolatile A term that describes a memory that can retain stored data when the power is removed
NOR gate A logic gate in which the output is LOW when any or all of the inputs are HIGH
NOT A basic logic operation that performs inversions
Numeric Related to numbers.

O

One-shot A monostable multivibrator.
OR A basic logic operation in which a true (HIGH) output occurs when one or more of the input conditions are true (HIGH)
OR gate A logic gate that produces a HIGH output when one or more inputs are HIGH
Oscillator An electronic circuit that is based on the principle of regenerative feedback and produces a repetitive output waveform; a signal source
Output The signal or line coming out of a circuit
Overflow The condition that occurs when the number of bits in a sum exceeds the number of bits in each of the numbers added

P

Period (T) The time required for a periodic waveform to repeat itself
Periodic Describes a waveform that repeats itself at a fixed interval
Pointer The contents of a register (or registers) that contain an address
Positive logic The system of representing a binary 1 with a HIGH and a binary 0 with a LOW

Preset An asynchronous input used to set a flip-flop (make the *Q* output *1*)

Primitive A basic logic element such as a gate or flip-flop, input/output pins, ground, and V_{cc}

Priority encoder An encoder in which only the highest value input digit is encoded and any other active input is ignored

Probe An accessory used to connect a voltage to the input of an oscilloscope or other instrument

Product term The Boolean product of two or more literals equivalent to an AND operation

Q

Queue A high-speed memory that stores instructions or data

Quotient The result of a division.

R

Race A condition in a logic network in which the difference in propagation times through two or more signal paths in the network can produce an erroneous output

Register A digital circuit capable of storing and shifting binary information; typically used as a temporary storage device

Register array A set of temporary storage locations within the microprocessor for keeping data and addresses that need to be accessed quickly by the program

S

Schematic (graphic) entry A method of placing a logic design into software using schematic symbols

Schottky A specific type of transistor-transistor logic circuit technology

Set-up time The time interval required for the control levels to be on the inputs to a digital circuit, such as a flip-flop, prior to the triggering edge of clock pulse

Shift To move binary data from stage to stage within a shift register or other storage device or to move binary data into or out of the device

Signal A type of VHDL object that holds data

Signal tracing A troubleshooting technique in which waveforms are observed in a step-by-step manner beginning at the input and working toward the output or vice versa. At each point the observed waveform is compared with the correct signal for that point

Sign bit The left-most bit of a binary number that designates whether the number is positive (0) or negative (1)
S-R flip-flop A SET-RESET flip-flop
Stage One storage element (flip-flop) in a register
State diagram A graphic depiction of a sequence of states or values
State machine A logic system exhibiting a sequence of states conditioned by internal logic and external inputs; any sequential circuit exhibiting a specified sequence of states
Storage The capability of a digital device to retain bits; the process of retaining digital data for later use
String A contiguous sequence of bytes or words
Subtracter A logic circuit used to subtract two binary numbers
Subtrahend The number that is being subtracted from the minuend
SUM The result when two or more numbers are added together
Sum-or-products (SOP) A form of Boolean expression that is basically the ORing of ANDed terms
Sum term The Boolean sum of two or more literals equivalent to an OR operation
Synchronous counter A type of counter in which each stage is clocked by the same pulse

T

Throughput The average speed with which a program is executed
Timing diagram A graph of digital waveforms showing the proper time relationship of two or more waveforms and how each waveform changes in relation to the others
Trigger A pulse used to initiate a change in the state of a logic circuit
Tristate A type of output in logic circuits that exhibits three states: HIGH, LOW, and HIGH-Z; also known as 3-state

U

Universal gate Either a NAND gate or a NOR gate. The term universal refers to the property of a gate that permits any logic function to be implemented by that gate or by a combination of gates of that kind
Universal shift register A register that has both serial and parallel input and output capability
Up/down counter A counter that can progress in either direction through a certain sequence

V

Variable symbol used to represent a logical quantity that can have a value of 1 or 0, usually designated by an italic letter
VHDL A standard hardware description language: IEEE Std. 1076-1993

W

Weight The value of a digit in a number based on its position in the number
Word A complete unit of binary data
Word capacity The number of words that a memory can store
Word length The number of bits in a word

References

1. V.C. Hamacher, Z.G. Vranesic, S.G. Zaky, *Computer Organization*, 5th edn. (McGraw-Hill, New York, 2002)
2. D.A. Patterson, J.L. Hennessy, *Computer Organization and Design: The Hardware/Software Interface*, 2nd edn. (Morgan Kaufmann, San Francisco, 1998)
3. D.D. Gajski, *Principles of Digital Design* (Prentice-Hall, Upper Saddle River, 1997)
4. M.M. Mano, C.R. Kime, *Logic and Computer Design Fundamentals* (Prentice-Hall, Upper Saddle River, 1997)
5. T.L. Floyd, *Digital Fundamentals* (Pearson Education International, Upper Saddle River, 2003)
6. R.J. Tocci, N.S. Widmer, *Digital Systems. Principles and Applications* (Prentice-Hall, Upper Saddle River, 2003)
7. J.P. Daniels, *Digital Design from Zero to One* (Wiley, New York, 1996)
8. V.P. Nelson, H.T. Nagle, B.D. Carroll, J.D. Irwin, *Digital Logic Circuit Analysis and Design* (Prentice-Hall, Englewood Cliffs, 1995)
9. R.H. Katz, *Contemporary Logic Design* (Benjamin/Cummings, Redwood City, 1994)
10. J.P. Hayes, *Introduction to Logic Design* (Addison-Wesley, Reading, 1993)
11. C.H. Roth Jr., *Fundamentals of Logic Design*, 4th edn. (West, St. Paul, 1993)
12. J.F. Wakerly, *Digital Design Principles and Practices* (Prentice-Hall, Englewood Cliffs, 1990)
13. R.M. Bertrand, *Programmable Controller Circuits* (Delmar Publishers, International Thomson Publishing, Inc., Albany, 1996)
14. B. George, *An Investigation of the Laws of Thought* (Project Gutenberg, 2005), www.gutenberg.net
15. K. Breeding, *Digital Design Fundamentals* (Prentice-Hall, Englewood Cliffs, 1989)
16. W.-K. Chen, *Logic Design* (CRC Press, Boca Raton, 2003)
17. E.D. Fabricus, *Modern Digital Design and Switching Theory* (CRC Press, Boca Raton, 1992)
18. F.D. Petruzella, *Programmable Logic Controllers* (McGraw-Hill, New York, 2005)
19. J. Daintith, Karnaugh map. A dictionary of computing (2004), Retrieved 25 September 2012 from Encyclopedia.com: http://www.encyclopedia.com/doc/1O11-Karnaughmap.html
20. E.V. Huntington, New sets of independent postulates for the algebra of logic, with special reference to Whitehead and Russell's Principia Mathematica. Trans. Amer. Math. Soc. 35, 274–304 (1933)

B. Borowik et al., *Theory of Digital Automata*, Intelligent Systems, Control and Automation: Science and Engineering 63, DOI 10.1007/978-94-007-5228-3,

Subject Index

B. Borowik et al., *Theory of Digital Automata*, Intelligent Systems, Control and Automation: Science and Engineering 63, DOI 10.1007/978-94-007-5228-3,

MIX
Papier aus verantwortungsvollen Quellen
Paper from responsible sources
FSC® C105338
FSC
www.fsc.org

If you have any concerns about our products,
you can contact us on
ProductSafety@springernature.com

In case Publisher is established outside the EU,
the EU authorized representative is:
Springer Nature Customer Service Center GmbH
Europaplatz 3, 69115 Heidelberg, Germany

Printed by Libri Plureos GmbH
in Hamburg, Germany